现代景观设计
Modern Landscape Design

法国 亦西文化 ICI Consultants 编
米歇尔·拉辛 Michel RACINE 前言
邵雪梅 陈庶 简嘉玲 译

辽宁科学技术出版社

现代景观设计 Modern Landscape Design

本图书为法国亦西文化公司(ICI Consultants/ICI Interface)的原创作品，原版为法英文双语版。
This work is created by ICI Consultants/ICI Interface, in an original French-English bilingual version.

法国亦西文化 ICI Consultants 策划编辑

总企划 Direction：简嘉玲 Chia-Ling CHIEN
协调编辑 Editorial Coordination：尼古拉·布里左 Nicolas BRIZAULT
英文翻译 English Translation：艾莉森·库里佛尔 Alison CULLIFORD
中文翻译 Chinese Translation：陈庶 Shu CHEN, 邵雪梅 Xue-Mei SHAO, 简嘉玲 Chia-Ling CHIEN
中文校阅 Chinese Proofreading：简嘉玲 Chia-Ling CHIEN
版式设计 Graphic Design & Layout：维建·诺黑 Wijane NOREE

Modern Landscape Design
现代景观设计

法国亦西文化 ICI Consultants 编
邵雪梅 陈庶 简嘉玲译

辽宁科学技术出版社

目录
Contents

景观与大型基地
Landscape and Site

012 前言
Preface

020 贝尔湖水滨公园
法国 贝尔莱唐
Berre Lakeside Park
Berre l'Étang, France

026 图尔河河岸公园
法国 塞尔奈
Thur Riverside Park
Cernay, France

032 圣杰克公园
法国 圣杰克-德-拉-兰德
Saint-Jacques Park
Saint-Jacques-de-la-Lande, France

038 露天运动游憩区
法国 托尔西
Sports and Leisure Park
Torcy, France

042 安洁丽克公园
法国 波尔多
Angéliques Park
Bordeaux, France

046 阿莱西亚博物馆公园
法国 阿利兹-圣-莱纳
Alésia MuseoPark
Alise-Sainte-Reine, France

050 维祖尔湖游憩区
法国 维祖尔
Vesoul Lake Leisure Park
Vesoul, France

054 阿里·帕鲁萨公园
俄罗斯 沃罗涅日
Alye Parusa Park
Voronezh, Russia

058 松林下的步道
法国 卡尔维
A Path through the Pine Forest
Calvi, France

064 马恩河岸
法国 托尔西
Banks of the Marne
Torcy, France

070 清翠园 – 相移公园
中国 台湾 台中
Eco Park – Phase Shift Park
Taichung, Taiwan

076 朗斯罗浮宫公园
法国 朗斯
Louvre-Lens Museum Park
Lens, France

082 巴尔河生态公园
法国 安格雷
Barre Ecological Park
Anglet, France

088 湿地公园 – 蒙多德汉航空中心
法国 图卢兹
Wetland Park – Montaudran Aerospace
Toulouse, France

公园与公共性花园
Public Parks and Gardens

094
加尔贝尔花园
法国 安纳西
Galbert Garden
Annecy, France

100
黑牛商业中心屋顶花园
法国 阿尔卡伊
Vache Noire Roof Garden
Arcueil, France

104
塞尔日·盖恩斯伯花园
法国 巴黎
Serge Gainsbourg Garden
Paris, France

110
风神花园
法国 巴黎
Zephyr Garden
Paris, France

116
塞甘岛预示花园
法国 布洛涅-比扬古
Seguin Island Prefiguration Garden
Boulogne-Billancourt, France

122
普朗诗岛屿溢洪道公园
法国 勒芒
Planches Island Park
Le Mans, France

126
马尔贝街区公园散步道
法国 马孔
Marbé Neighbourhood Park-Promenade
Mâcon, France

134
东印度公园
法国 洛里昂
Park of the West Indies
Lorient, France

138
彼岸
加拿大 魁北克
The Other Bank
Québec, Canada

144
Spot5花园 / 山水意象
中国 西安
Spot5 / Shanshui "Mountain Water"
Xi'an, China

148
游戏场公园
法国 埃尔蒙
Playground Park
Ermont, France

150
马丁·路德·金公园
法国 巴黎
Martin Luther King Park
Paris, France

156
伯恩生态街区公园
法国 格勒诺布尔
Park for Bonne Eco-neighbourhood
Grenoble, France

162
桑特里居民花园
法国 塞纳河畔桑特里
Saintryiens Garden
Saintry-sur-Seine, France

168
克雷泰伊湖岬角
法国 克雷泰伊
Créteil Lake Point
Créteil, France

174
东区城市公园
英国 伯明翰
Eastside City Park
Birmingham, United Kingdom

180
圣彼得堡动物园
俄罗斯 圣彼得堡
Saint-Petersburg Zoo
Saint-Petersburg, Russia

184
艾夫兰山公园
法国 蒙泰夫兰
Mount Évrin Park
Montévrain, France

城市公共空间
Urban Public Spaces

192
保罗·格里莫花园广场
法国 安纳西
Paul Grimault Esplanade-Garden
Annecy, France

198
"林中居所" – 欧洲里尔计划 2
法国 里尔
"Inhabited Wood" – Euralille 2
Lille, France

202
城堡广场
德国 卡尔斯鲁厄
Schlossplatz
Karlsruhe, Germany

206
夸斯兰停车场
法国 梅斯
Coislin Car Park
Metz, France

210
拉马丁广场
法国 马孔
Lamartine Esplanade
Mâcon, France

216
天顶音乐厅外围空间
法国 斯特拉斯堡
Zénith Concert Hall Public Realm
Strasbourg, France

224
瓦兹河岸
法国 蓬图瓦兹
Quays of the Oise
Pontoise, France

228
欧洲南特火车站 / 马拉科夫
法国 南特
Euronantes Station/Malakoff
Nantes, France

234
三河林荫道
法国 斯丹
Three Rivers Mall
Stains, France

240
阿里斯蒂德·白里安林荫大道
法国 努瓦西-勒-格朗
Aristide Briand Avenue
Noisy-le-Grand, France

246
上德勒河岸生态街区
法国 里尔
Banks of the High Deûle Eco-neighbourhood
Lille, France

254
戴高乐广场与圣瓦斯特广场
法国 阿尔芒蒂耶尔
Charles de Gaulle & Saint Vaast Square
Armentières, France

260
波提耶尔-申内生态街区
法国 南特
Bottière Chênaie Eco-neighbourhood
Nantes, France

272
奥奈协议开发区
法国 奥奈丛林
Aulnes ZAC
Aulnay-sous-Bois, France

276
伊莲D与查尔斯A·萨蒙斯公园
美国 达拉斯
Elaine D. & Charles A. Sammons Park
Dallas, United States of America

282
奥斯特里茨广场
法国 斯特拉斯堡
Austerlitz Square
Strasbourg, France

288
施托克教士广场
法国 沙特尔
Abbé Stock Square
Chartres, France

292
塞纳河岸街区
法国 巴黎
Front de Seine Neighbourhood
Paris, France

296
罗纳河岸
法国 里昂
Banks of the Rhône
Lyon, France

312
帕若尔协商开发区
法国 巴黎
Pajol ZAC
Paris, France

316
勒阿弗尔城市入口
法国 勒阿弗尔
Le Havre City Entrance
Le Havre, France

324
奥蒂埃林荫大道
法国 塞尔日-蓬图瓦兹
Hautil Boulevard
Cergy-Pontoise, France

330
马泽尔广场
法国 梅斯
Mazelle Square
Metz, France

334
弗朗索瓦·密特朗林荫道
法国 雷恩
François Mitterrand Mall
Rennes, France

338
四季广场
法国 拉德芳斯
Seasons Square
La Défense, France

342
帕甬河畔景观步道
法国 尼斯
Paillon Landscape Walks
Nice, France

346
拱廊广场
法国 埃佩尔奈
Arcades Square
Épernay, France

350
贝瑞广场
法国 亚眠
Perret Square
Amiens, France

354
兰斯火车站站前广场
法国 兰斯
Reims Station Forecourt
Reims, France

358
贝里运河
法国 谢尔省
Berry Canal
Department of Cher, France

362
圣康坦-伊夫林城市中心区
法国 圣康坦-伊夫林
Saint-Quentin-en-Yvelines Central Urban Cluster
Saint-Quentin-en-Yvelines, France

368
解放广场
法国 特鲁瓦
Liberation Square
Troyes, France

372
共和广场
法国 巴黎
Place de la République
Paris, France

376
奥斯特利兹堤岸
法国 巴黎
Austerlitz Quays
Paris, France

私人花园与绿化空间
Private Gardens and Landscape Spaces

384
格雷恩城堡绿色剧场
法国 格雷恩
Open-Air Theatre – Château de Grâne
Grâne, France

390
达索系统企业园区
法国 维利兹-维拉库布莱
Dassault Systèmes Campus
Vélizy, France

396
法国国家地理信息中心与气象局花园
法国 圣蒙德
Garden of the IGN and Météo France
Saint-Mandé, France

398
维利兹大道企业园区
法国 维利兹-维拉库布莱
Vélizy Way Park
Vélizy-Villacoublay, France

402
维朗德丽城堡中的两个花园
法国 维朗德丽
Two gardens of Château de Villandry
Villandry, France

406
庞格城堡花园
法国 庞格
Gardens of Château de Pange
Pange, France

410
绿色伊甸园
法国 巴黎
"Eden Bio" Urban Hamlet
Paris, France

416
伯纳尔博物馆
法国 勒卡内
Bonnard Museum Gardens
Le Cannet, France

422
小姐妹教会花园
法国 土伦
Garden of Little Sisters of the Poor
Toulon, France

428
阿朗松大学校园
法国 阿朗松
Alençon University Campus
Alençon, France

432
法国大使馆花园
中国 北京
Gardens of the French Embassy
Beijing, China

436
卡尔普·迪尔木大厦花园
法国 拉德芳斯
Garden of Carpe Diem Tower
La Défense, France

442
陶艺花园
法国 萨尔格米纳
Potters' gardens
Sarreguemines, France

450
奥德赛花园2000
法国 南泰尔
Odyssey 2000 Gardens
Nanterre, France

456
建筑物平台企业花园
法国 欧贝维利耶
Gardens for La Plateforme du Bâtiment
Aubervilliers, France

460
"杰克逊·波洛克"办公楼花园
法国 圣丹尼
"Jackson Pollock" Office Garden
Saint-Denis, France

464
皮埃尔-乔尔·邦德建筑职业高中
法国 里永-沃尔维克
Pierre-Joël Bonté School for Building Skills
Riom-Volvic, France

468
宏格耶大学校园
法国 图卢兹
Rangueil Campus
Toulouse, France

472
附录
设计师索引
Annex
The designers

前言
Preface

Michel RACINE 米歇尔·拉辛

米歇尔·拉辛：法国景观建筑师、文化景观ICOMOS-IFLA国际科学委员会具投票权会员、凡尔赛国立景观设计高等学院教师、《中国景观建筑师》（CLA）编辑委员，同时也是法国众多有关花园、景观和景观设计师的图书作者。

Michel Racine: landscape architect, voting member at the International Scientific Committee ICOMOS-IFLA on Cultural Landscape, teacher at the National School of Landscape Architecture in Versailles, member of the editorial committee of *Chinese Landscape Architecture* (CLA), author of several books on gardens, landscape architecture and landscape architects.

世界重新发牌

在这个世界急剧转变的时刻,法国的景观设计师们要向我们说些什么?他们的文化和专业知识如何在地方尺度和全球尺度上回应当今社会、生态和景观层面的挑战?他们如何参与在景观、建筑、造型艺术等领域所展开的创作浪潮?他们是否拥有独特的见解与手法?

这本书[1]企图带领读者一览当今法国景观设计的面貌,为这些景观方案定位。它也为与景观领域相关的政治决策者和专业人士提供立即而有效的参考,为那些有意无意间参与了景观形成的人们展现成果,为时下的景观创作留下历史见证。

在这个各个领域皆面临重大动荡的时期,不论是在科学、产业层面的动荡,或者在思想、人与物品的运输层面的动荡,我们从未如此需要有人来仔细反省我们生活环境的变化,继而为其进行构思——从小住宅到大景观,从与居民的协商开始,直到设计、施工和日常维护管理,都重新赋予它们意义、使它们能够在时间中继续发展演变。

自从人类从太空中发现地球的影像、从花园土壤里认识到微生物组织的丰富性以来,人们已经无法忍受那些位于世界金字塔顶端的人们一直到不久前还维持着的生活方式。面对各种尺度的环境挑战,倘若人们还在伟大建筑师的神话形象面前退缩不前,或者将景观师的角色仅仅限定于是赋予方案形式的人,那这些想法就再天真不过了。我们从不曾像今天一样对景观创造者有着如此深厚的期待。他的方案必须将与基地和其未来演变有关的所有因素纳入考量,将社会的、历史的、生态的、经济的条件资料融入构思之中。他必须与居民以及所有与基地相关的人们进行沟通,以便

New world order

At a time when the world is changing ever faster, what do French landscape architects have to say? What is the essence of their culture and how do they use their know-how to tackle the social, ecological and landscape challenges of our time – challenges that are both on a local and global scale? What part do they play in the larger creative movements in landscape, architecture and design? Do they have a particular approach of their own?

This book[1] aims to give an overview of contemporary French landscape architecture, to position it, and to offer an immediate and useful resource for policy-makers and professionals involved in this field, as well as for everyone who plays a part, consciously or not, in the history of today's landscape.

In a period characterised by great upheavals in every field, from sciences to business and to transmission of ideas, people and goods, we have an even greater need for designers who can think through the metamorphoses of our daily habitat. We need professionals who can make the leap from the home to the larger landscape, who can give meaning and enable it to work with time, carrying a project through, from hearings with the inhabitants, the conception in the studio, the construction work execution to the means of everyday management.

Not so long ago Man was seen at the top of the pyramid of the living world, but with the discovery of images of Earth from outer space, and those showing the richness of the micro-organisms of the earth in our gardens, this no longer makes sense. Faced with challenges on such different scales it would be naive to give in to the image of the great Architect or to limit the role of the landscape architect as a simple giver of form. Today, more than ever, we expect a lot from our landscape creators. Their projects must integrate a variety of social, historical, ecological and economic factors concerning a site and its possibilities for evolving. Form comes from an idea built on a

创造一个有意义的空间形态和场所，通过环境氛围、材质、颜色和气味的品质来引发人们的感官知觉，总结出一个与男男女女、老老少少、健全者和残障人士的身体与心灵都相关的综合方案。此乃为景观师的雄厚企图和愿望。

新期望、新计划与新诠释

荒地，废弃的港口、机场和铁路用地，河流与湖泊的岸边…… 在大城市以及其邻近郊区，许多大型空间被释放出来、突然间处于新建街区的一旁并且受到诸多质疑。居民们期望用这些空地来满足新的使用需求，政治人物或业主期望用它们来为一个充斥着水泥沥青的街区提供自然空间、散步道、城市公园、广场、花园。面对每一个项目，景观师的设计团队往往置身于复杂的课题当中，必须经过漫长的协议过程来将这些考量因素转化为具体的景观方案。

我们距离20世纪中期的"预防性绿化空间"[2]的规划已经相当遥远，当时的娱乐游憩场和运动设施的规划仅仅是让景观师们来为工程师与建筑师所留下的空间种种花草树木，如今景观师则必须全程参与整个方案的构思过程。人们寻找场所的记忆并且强调生物多样性，着重对自然生态环境和生态系统的认识，也重视可持续发展。生态与环保概念以及其教育实践从此被纳入于所有的新计划当中，甚至出现在新景观场所的命名当中，例如"生态公园"或"生态街区"[3]，大自然在在被邀请进到城市的中心。于是，景观师与城市规划师和建筑师比往常更加频繁地并肩作战，但愿这些合作能够持续发展下去！人们越来越乐于进行一些大自然中的享乐活动：散步、躺卧在草地上、野餐、与人交流、观赏景观，并且追求通过空间组织与环境互动所带来的感官经验和美感愉悦。人们正在开创新的居住环境。

real dialogue with the residents and the site stakeholders. They give sense to the place, and call out to sensations by working on the quality of environments, materials, colours, and scents. It is a common core implicating men, women, children, old or disabled people – body and soul. That is the aim.

New expectations, new programmes, new interpretations

Brownfields, defunct harbours, airports and railway land, riverbanks and lakes… In towns and their surroundings, large spaces are being freed and suddenly finding themselves near new neighbourhoods. They open themselves up to questioning, to the residents' desire to invest them with new uses. They give policy-makers or land owners the possibility of offering the residents of "concrete-asphalt" neighbourhoods access to natural space, to walks, to urban parks that they can really make their own, to a playground or garden. Each time, teams of landscape architects find themselves confronted with complex problems that, through a long process of consultation, must be translated into a landscape project.

We have moved on from the mid-20th century's programmes of "therapeutic green spaces"[2], of recreational parks or sports facilities where landscape architects were simply called on to plant spaces left behind after the engineers and architects had done their work. The landscape architect now has a role throughout the whole design process. We search for the memory of places, for biodiversity, for the discovery of the natural environment, for ecosystems and sustainable development. Ecology and the ways of teaching it now have a strong place in all programmes, even down to the very name of the new landscapes that are now called "ecological parks" or "eco-neighbourhoods"[3]. Nature is invited into the heart of the city. Landscape architects work more often with town planners and architects. Let's hope that these long-awaited collaborations continue to develop! We open up ourselves to more sensual ways of enjoying the landscape: walking, lying in the grass, picnicking, meeting people, contemplating the view – and to searching for an aesthetic sense of pleasure provoked by the architect of the space and his relationship to the cosmos. We invent a new habitat.

近30年来法国民众对花园的热衷使得如今"花园"的概念在环境规划中强行复出。经过景观师以当代眼光的检视，在生物多样化的要求以及民众参与之下，现今的花园展现出多种形态：野餐绿地、生态教学花园、社团实验性花园……花园一如所有绿化空间，其应用甚至被延伸到水泥空间里：楼板平台上的花园也能提供"一系列古玫瑰，令人联想起阿尔卑斯山的野生玫瑰"[4]。接下来必须注意的是重新赋予园丁发挥其专精园艺的空间，并且避免让绿化空间的规划被简化为单纯的绿色设计。

同样地，在城市广场、林荫道上，在沿着河岸、运河畔或者松林下设置的散步道里，城市景观艺术的贡献也都获得了肯定。景观师尝试着在所有的尺度上"确保行进的连续性"，并且"重新缝合"、"让人接近"和连接不同街区，他们也致力于"钉扣住"城市与郊区、城市与自然景观、水景、湿地环境、河流、池塘……为它们"重新编织关系"。

景观师重新创造了一种绿色城市规划，一种"可辨识"的空间，在其中，所有景观的展现都归于人们的分享。

看似一家人

在不可避免的全球化的环境背景下，以及在民众对其日常生活环境品质的提升要求下，景观与园林的规划设计者所背负的责任与时俱增。法国的设计师们以什么样的专业态度来担扛这份责任？这本书所呈现的两代景观师的最新作品，显现出一些值得强调的特质，特别是他们同时处理大小不同尺度项目的从容能力：从花园、广场、林荫大道，直到大地景观。这个处理景观项目的共通文化，尽管各人依照各自性格而有不同发挥，但仍然显示出了一个培育环境所带来的影响。这些景观师中许多出自凡尔赛国立景观设计高等学院，而不论师出何门，他们都承袭了法国至少四个世纪以来在"结合城市空间与园林"这个领域的创新经验，不论是形态规律的、所谓的"法式园林"，

The idea of "garden" is back, stronger than ever, reinvigorated by the French public's passion for gardens in the past thirty years. Revisited by landscape architects with a contemporary approach that considers biodiversity and the involvement of citizens, a garden takes all kinds of forms: meadows for picnics, gardens of ecological discovery, experimental community gardens… The garden is justified by the idea of green space and moves right up to, and even onto, concrete spaces: a garden on a hard surface plaza can offer "a collection of old rose varieties, evoking the wild roses of the Alps"[4]. What is needed now is to give the gardener back his/her place and, beyond the poetic words, not to reduce the evolution of green space to a simple "green design".

It's also an art of urban landscape that asserts itself in programmes for urban playgrounds and avenues, for walks by the river, the canal, on the quays, under the pines. In all aspects, the landscape architect aims to "ensure continuity", to "sew together again", to "give access" and link neighbourhoods to each other, to "staple", "reweave the links" between town and suburb, town and natural landscape, waterside landscape, wetland environments, rivers and lakes.

He reinvents town planning in a green form, creating understandable spaces to be shared.

A family likeness

In the unavoidable context of globalisation and the increase in public expectations about the quality of their everyday space, landscape and garden designers have been entrusted with growing responsibilities. How do French landscape architects rise to this challenge? The overview of recent projects by two generations of landscape architects presented here highlights specific strengths that deserve to be emphasised – in particular their ease when simultaneously treating small and large scales, from the garden, the playground or the avenue to the larger landscape. This common culture of landscape project, adapted to each personality, reveals the influence of an education – for many that of the National School of Landscape Architecture in Versailles – but also of a heritage. For at least four centuries the French have been innovating in ways of bringing urban space and garden together, both in the ordinate gardens known as "jardins

还是景观式园林。这是为什么这些景观师们能够关注地理环境、地块文脉、结构性的植树排列、树林围篱网络、景观与建筑和基础设施的对话，具备一种从花园到土地的景观思考方式。这些21世纪初的法国景观师延续了先前几代园林设计师与城市规划师所传承下来的传统，德扎利埃·达让维尔在1709年写下："园林中的通道有如城市中的街道"⁵。这个想法被20世纪初的景观师和城市规划师反过来运用，犹如J.C.N.弗里斯蒂埃和亨利·普斯特所说的："城市中的街道有如园林中的通道"。我们可以猜想，这些新散步道的设计者对于路易十五时期所开设的城市散步道或者安德烈·勒诺特尔在圣日耳曼昂莱居高临下、面对着塞纳河所建造的著名散步平台仍然保有鲜明记忆。这些优秀的景观师们对于注重场所氛围与空间材质的这种敏锐而感性的构思方式也同样产生认同，他们寻求场所的诗意，希望创造出"拥有血肉与灵魂的景观"⁶。

园艺景观师

随着人们对于城市中各种尺度的花园的关注，15年来，我们不断看到从园艺培训或者实地培训而出来的花园设计者。相对于大部分历史上的著名景观设计师而言，这也是另一种来自园艺世家的传统。最有名的例子则是安德烈·勒诺特尔，其父亲与祖父都是园艺工匠，他们因为受到伟大艺术家的熏陶而逐渐学会将空间转化为景观。带着诗意的眼光、充实的技术、对植物的细致鉴赏能力以及能够确保持续性的生态创作手法，这些景观园艺师为私人花园或历史性园林带来了清新面貌与多样色彩。这些专精园艺的花园创作者也提醒了一个事实：花园经常成为在其他相近领域（例如植物学或艺术创作）具有相当经验者尽情发挥的天地。

随着我们对环境认知的不同、我们眼光的转变、我们与自然关系的演化、我们使用城市、乡间与土地方式的变更，花园或景观创作的专业实践也一直在演变当中。

à la française" and in landscaped gardens. This is what gives this attention to geography, to the division into smaller spaces, to the structuring lines, to a hedgerow network, to the dialogue with architecture and infrastructure, to a French way of thinking about landscape – from garden to regional area. At the beginning of the 21st century, French landscape architects are perpetuating a real tradition inherited from generations of garden designers and town planners. The idea expressed in 1709 by Dezallier d'Argenville that "Avenues in a garden are like the streets of a city"[5] was mirrored by French landscape architects and town planners at the beginning of the 20th century, like Jean Claude Nicolas Forestier and Henri Prost, who said "the streets of a city are like the avenues in gardens". The creators of new walks certainly have in mind the famous walled walk overlooking the Seine created by André Le Nôtre in Saint-Germain-en-Laye or the urban walks developed under Louis XV. But the best of these landscape architects also identify with an approach that is sensitive and sensually guided by atmospheres and materials, that searches for a poetry of place, for a project that creates "landscapes that have body and soul"[6].

The ascent of the landscape gardener

With growing interest in city gardens on every scale, the past decade has seen a rise in designers whose training is in gardening. This is also a tradition, as the majority of the great landscape architects came from dynasties of gardeners – the most famous example being André Le Nôtre, a gardener, son and grandson of a gardener, who, through meeting great artists learnt to read space as a landscape. With a poetic approach, a great know-how, refined plant palettes and an ecological approach that ensures the longevity of their creations, these landscape architects, who are also gardeners, bring freshness and colour to private projects and historic gardens. This reminds us that the garden has always been a place where those who already have experience in neighbouring domains, such as plants and arts, flourish as new "garden creators", undertaking commissions after training on specialised courses[7].

The practice of the profession of garden and landscape designer has never stopped evolving along with our knowledge of habitat,

随着时代变迁，这些景观创作者被称为园丁、园艺师、园林设计师、建筑师、景观师……然而，在这个不断变化的专业中，不论使用什么语言，以单一称谓来定义这个职业多少都缺乏了全面性的诠释。在国际领域上，景观建筑师（Landscape architect）成为人们共同认知的称谓，然而它却很难涵括这个专业所有的领域：园艺、工程、生态、调解、灯光……18世纪英国园林大师胡弗莱·雷普顿则自称为景观园艺师（landscape gardener）[8]。在设计理念、花园与景观文化不断变化的国际环境中，在各种语言用词和翻译也不断演变的背景中，英文design一词的双重含义[9]以及景观设计师（landscape designer）的用词暂时成为我们的一种支援。然而，既然花园的时代又前来滋养景观的时代，倘若我们希望将对花园的热情与对景观的喜爱结合起来，那么景观园艺师的概念也许值得获得新生[10]。

changes in the way we see the world, our relationship with nature, our way of inhabiting cities, countryside, or regions.

According to the time they lived or live in, these creators have been variously called gardeners, architects, landscape architects, gardenists. It will always be an over-simplification to try to define this profession in perpetual movement with a single word, whatever language we speak. Landscape architect, the internationally recognised term, hardly covers all the skills that are necessary today: gardening, engineering, ecology, mediation, lighting design… Humphrey Repton's term of landscape gardener[8] deserved to be translated as "jardinier du paysage". In the international context of the evolution of ideas and cultures of garden and landscapes and of their translation into different languages, English comes to our rescue with the double meaning of design[9] and the term landscape designer. But since the age of gardens has come back to nourish the age of landscape, if we need to link our twin loves of landscape and garden, the attractive idea of landscape gardener might deserve a new life[10].

注释：
1. 本书和其前身《法国景观设计》（法英文版《Expression paysagère, Création française / French landscape design》，ICI Interface出版社，巴黎，2007年12月。中文版由辽宁科学技术出版社于2007年10月出版），两本具有类似效用。
2. 根据阿兰·普沃的说法。
3. 里尔市的上德勒河岸生态街区、南特市的波提耶尔-申内生态街区、城市自然公园。
4. 安纳西市的加尔贝尔花园。
5. 安东尼-约瑟芬·德扎利埃·达让维尔，《造园的理论与实践》，1709和1747；Actes Sud出版社于2003年再版。
6. 皮埃尔·桑索，法国哲学家、社会学家和作家。
7. 凡尔赛国立景观设计高等学院所开设的培训课程《景观中的花园构思》（Conception de jardin dans le paysage / Garden and landscape design）。
8. 约翰·迪克松·亨特采用"景观花园设计师"来称呼威廉·肯特。
9. 绘图和设计。
10. 并获得一个适当的翻译……

Notes:
1. This second book as the first one, "Expression paysagère, Création française / French landscape design", ICI Interface, Paris, december 2007.
2. Allain Provost's expression.
3. Banks of the High Deûle Eco-neighbourhood, Lille; Bottière Chênaie Eco-neighbourhood, Nantes; Natural Urban Park.
4. Galbert Garden, Annecy.
5. Antoine-Joseph Dezallier d'Argenville, "La théorie et la pratique du jardinage", 1709, 1747. Reprint, Actes Sud, 2003.
6. Pierre Sansot.
7. Diploma of "Conception de jardin dans le paysage / Garden and landscape design", National School of Landscape Architecture in Versailles.
8. John Dixon Hunt used the term "landscape garden designer" for William Kent.
9. A drawing, plan or motif, and an aim or intention.
10. Along with a good French translation…

01

Landscape and Site

景观与大型基地

贝尔湖水滨公园
Berre Lakeside Park
AGENCE APS

地点：法国贝尔莱唐
完工日期：2008-2011
面积：50 ha
业主：贝尔莱唐镇政府
合作设计师：Cabinet Merlin
照片版权：Agence APS

Location: Berre l'Étang, France
Completion date: 2008-2011
Area: 50 ha
Client: Berre l'Étang Town Council
Co-projet manager: Cabinet Merlin
Photo credits: Agence APS

01. 总体规划平面配置图
02. 公园为城市与贝尔湖之间建立起联系
03. 观景台小丘成为散步者的视觉焦点
04. 从观景台小丘上看向城市、公园与湖泊

01. Master plan
02. The park restores the links between the town and the lake
03. The lookout hill with its belvedere offers rewards for the walker
04. View of the town, the park and the lake from the belvedere

贝尔莱唐（贝尔湖）的海军航空基地使得处于城市中心附近的一大部分湖岸长久以来都是无法接近的。而如今，马赛/普罗旺斯机场被设置于对岸，与城市隔岸相望。这块在城市发展上具有决策性的土地因而得以被释放出来，并由市政府征收，促使了湖岸的变更计划。一片占地50公顷、沿着两公里长水岸延伸的线性公园将在此实现。

作为公园设计方的APS事务所也被委托负责基地的整体规划，在保证公园延续性的同时让城市向水面延伸。一条新的城市林荫大道把城市中心从繁重的公路交通中解放出来。围绕着这条整合性轴线，一系列新的横贯机动车道和人行林荫路重新建立了城市与水的关联。

在这个大型"都市自然公园"中，三个序列空间由西侧到东侧逐渐从最城市化的场所转变为最自然的场所，既展示出现有基地内的历史痕迹，又发掘出与其相对应的发展潜力。"大草场码头"是位于湖滨、布满树荫的宽阔场地，成为名副其实的城市绿肺，它背靠着小港口，是湖边的散步道的出发地。"棕榈树广场"是一大片石灰石铺就的梯形空地，作为维克多·雨果大街的出口，它是城市中心的自然延伸。在基地东侧则发展出"湖滨公园"。一条与湖岸平行、植物繁茂的沟渠从旧水上飞机库旁一直延伸到一座作为观景台的小丘。一条螺旋状的缓坡领人通达小丘顶点，观景台以其12米的高度供人们远眺，展现出城市、公园、池塘和景观视野之间的关联。这个项目赢得了由里昂市颁发的2009年"城市之光"大奖里的公共空间奖项。

For a long time Berre l'Étang naval air station prevented access to the banks of the brine lake at town centre level. Today Marseille Provence airport is situated on the opposite bank of the lake, facing the town. The liberated site, and the acquisition by the town of this strategic area for its development, allowed for the development of the bank with the arrival of a long, linear park covering 50 hectares, along more than 2 kilometres of waterside.

An urban masterplan entrusted to the park's designer, APS agency, frames the town's extension towards the lake while preserving the continuity of the park. A new urban boulevard frees up the town centre from road traffic. Around this federating axis a new network of side roads and pedestrian paths reaffirms the relationship between the city and the lake.

In this great "urban natural park" three sequences are played out from the most urban in the west to the most natural in the east, revealing the existing lines and the potentialities. A wide lawn scattered with shade-giving trees, the "great meadow mole" provides a lung for the city. The lakeside promenade backing onto the little port begins here. "The palm esplanade" is a wide trapezium of limestone paving and the natural extension of the city centre from Victor Hugo Boulevard. The "bankside park" advances further to the east. Parallel to the lake, a long planted canal stretches between the old seaplane hangar and the lookout hilltop, where, from 12 metres up, a belvedere captures the view. Approached via a soft spiral ramp, it reveals the links that unite the city, the park, the plake and its views. Thanks to the work of L'Atelier Lumière, the project received the Lumiville public space trophy 2009 from the city of Lyon.

05. 架在水面上的浮式码头沿路点缀着散步道
06. 棕榈树广场是城市的"自然"延伸
07. 一片木质圆形地板提供人们一个面向湖泊的绝妙"布景"
05. Small pontoons on stilts punctuate the promenade
06. The palm esplanade is the "natural" extension of the town centre
07. A huge circular deck offers a spectacular view over the lake

01 景观与大型基地 Landscape and site

08. 基地内的老铁轨痕迹伴随着植物沟道
09. 条形分布的蓝色羊茅装点着公园的某一个入口
10. 游戏场位于市中心、学校与公园之间

08. Traces of the old railway flank a long planted bed
09. Lines of blue fescue criss-cross one of the park entrances
10. The children's playground is at the interface between the town centre, the schools and the park

01 景观与大型基地 Landscape and site

Thur Riverside Park
图尔河河岸公园
ATELIER VILLES & PAYSAGES

地点：法国塞尔奈
完工日期：2008
面积：12 ha
业主：塞尔奈镇政府
合作设计师：Egis France
照片版权：Atelier Villes & Paysages (n°09, 10), Christophe Bourgeois (n°07, 08), Zeppeline (n°01, 03-05)

Location: Cernay, France
Completion date: 2008
Area: 12 ha
Client: Cernay Town Council
Co-project manager: Egis France
Photo credits: Atelier Villes & Paysages (n°09, 10), Christophe Bourgeois (n°07, 08), Zeppeline (n°01, 03-05)

01. 一系列的过桥带人通往可以吸收洪泛的花园岛屿
02. 总体平面配置图

01. A set of footbridges serves the islands of the floodable garden
02. Master plan

沿河两岸伸展的图尔河河岸公园面积有12公顷，基于对洪水泛滥的恐惧，这个位于莱茵河上游、拥有11000居民的塞尔奈城市长期以来背对着河流发展，而如今市政府把这个将土地切割开来、遭人仇视的元素转化为一个结构性元素，以对城市进行一个极具必要性的缝合整治。

此公园位于一个迅速发展的新兴街区当中，其方案必须达成多重目标：城市扩展、居住混合和对大自然的尊重。环境的可持续发展为民意代表所关注的重点，因此成为项目计划的基础元素。这个场所必须成为充满活力的城市公园，提供居民各种休憩和游戏的可能性，时而带领人们进入森林内部、来到林中空地，时而化身为游戏场所、花园、绿化阶梯等。

人们沿着散步路线陆续发现五个高达7米、由柳条编织成的"捕鱼篓"，它们令人回想起18世纪的凉亭与花园官邸，并成为塞尔奈公园的标志性物体，每一个"捕鱼篓"都配备着不同的设备，使它们各具特色并且在散步道中扮演不同角色，各自呈现"接待"、"地景"、"沙龙"、"水井"、"观景台"等主题。

The Thur riverside park extends over 12 hectares on the two sides of the river. Having long turned its back on the water course out of fear of its frequent flooding, the town of 11,000 inhabitants in the High Rhine has turned this hostile element, which broke up the land area, into a structuring element, which called for an act of urban repair.

Located in a new, booming neighbourhood, the park has several objectives: urban extension, mixed housing and respect for nature. Sustainable development was already a real preoccupation for the local governing bodies and made up a fundamental element in the programme. The place has thus been designed as a real urban dynamic park, offering many possibilities for relaxing and playing, whether in the under wood or in the clearings, whether through playgrounds, gardens, or natural terraces.

Along its itinerary, the walker discovers five *nasses*, a kind of "willow basket" 7 metres high. Symbolic objects for the Cernay park, they recall the kiosks and follies of the 18th century. Each is equipped with different elements that give it a particular identity and a role in the walk. Thus you find the themes of "welcome", "the tree", "living room", "well" and "look-out tower".

03. 这片城市中的大型林中空地成为人们休闲与放松的场所
04. 位于小型林中空地中的"沙龙"捕鱼篓
05. 河岸公园：从巴榭桥望向"接待"捕鱼篓

03. The large urban clearing, a place for leisure and relaxation
04. The "living room" *nasse* in the little clearing
05. The riverside park: the "welcome" nasse seen from Basset bridge

01 景观与大型基地 Landscape and site

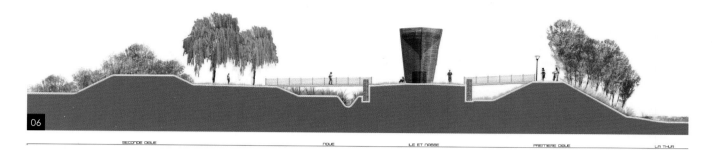

06. 柳树群和"水井"捕鱼篓区段剖面图
07. "沙龙"捕鱼篓是人们进行野餐和交流的场所
08. 河岸公园中的游戏场
09. 大型林中空地的石笼阶梯同时也作预防涨潮的装置
10. "日光浴场"

06. Section of the collection of willows and the "well" nasse
07. The "living room" nasse, a place for picnics and meeting people
08. The riverside park's children's playground
09. The gabion terraces of the large urban clearing reduce the likelihood of flooding
10. Sun-bathing area

01 景观与大型基地 Landscape and site

Saint-Jacques Park
圣杰克公园

ATELIER DE PAYSAGES BRUEL-DELMAR

地点：法国圣杰克-德-拉-兰德
完工日期：2007-2010
面积：40 ha
业主：圣杰克-德-拉-兰德镇政府
照片版权：Atelier de paysages Bruel-Delmar (n°02-08, 12-15), Antoine Guilhem Ducléon (n°10, 11), Christian Reland (n°09)

Location: Saint-Jacques-de-la-Lande, France
Completion date: 2007-2010
Area: 40 ha
Client: Saint-Jacques-de-la-Lande Town Council
Photo credits: Atelier de paysages Bruel-Delmar (n°02-08, 12-15), Antoine Guilhem Ducléon (n°10, 11), Christian Reland (n°09)

01. 总体平面配置图：生态公园是各式空间环境与各样生物活动的汇总
02. 人工芦苇湿地，同时具备生态与去污染功能
03. 在一个植物沙龙中享受阳光
01. Master plan: the ecological park is a mosaic of diversity
02. The artificial reed bed, ecology and remediation
03. Sun bathing in an open-air living room

圣杰克-德-拉-兰德公园与一个有关土地的、历史的、社会的以及政治的时间背景和环境背景关系紧密，这些元素构成了此城市规划项目的特殊性，并为它的更新提供了有利条件。

此市镇具有众多优点特征，而规划项目则建立在对这些特殊性的认知的基础上，并遵循三个基本概念来进行设计。首先，通过对重要景观元素的识别来重新建立公园与地理特征之间的关联，例如维兰河及其支流这样的基本元素。其次，通过对贺冬大道的重新整治来确保三个街区在象征性和功能性上的联结。最后，在城镇的肌理间，以公园或者散步道的形式来使每个街区都与普黑瓦雷的土地建立联系。

无论是从规划策略还是形式语汇上来讲，此公园从维兰河和布鲁森河开始就趋于朴实，逐渐融入到小树林围篱所形成的网状结构中，没有强制赋予景观任何额外的意象。它同时也确保了街区间东西向的连贯性。作为新的城市中心，公园里设置了休闲设施以满足居民的要求，并为他们提供了一块探索湿地生态环境的场所。不同种群的生物遍布在40公顷的公园用地上，并受到科学性的维护和管理，以保持多样混合的状态。这同时也成为此地区的识别特征，在保留了古人农耕活动的印记的同时，也重新获得天然的生态环境。

The park of Saint-Jacques-de-la-Lande forms part of an urban renewal project that unites land use, historical, sociological and political aims.

The urban project is based on a recognition of the specific qualities of the commune, following three principles. First of all, it is about rediscovering a relationship to the geography by identifying the main cohesions in the landscape, with the Vilaine and its tributaries as a cornerstone. Next, ensuring the symbolic and functional link between the three neighbourhoods through the reworking of the Redon road. Finally, connecting each neighbourhood with the land area of La Prévalaye through parks or walks that run throughout the urban fabric of the commune.

The Vilaine and Blosne rivers are the starting point for the park, both in terms of strategy and of vocabulary. It moulds itself to the landscape of small fields and hedgerows without superimposing a new image, and ensures the main east-west continuities between neighbourhoods. Forming a new heart for the town, it responds to the needs of the inhabitants by offering leisure facilities as well as a space for discovering the ecology linked to wetland environments. Covering 40 hectares, it embraces a diversity of environments, managed sustainably in order to preserve the mosaic that forms its identity while also allowing spontaneous ecological reconquest without erasing the memory of ancestral cultural practices.

04. 位于橡树林脚下的沉淀池
05. 伸展于鸢尾池塘上的浮桥
06. 青少年学子对生态环境进行观察
07. 在树林围绕的地块中新开辟的芦苇沼泽
08. 衡量生态动力而作的空间装置

04. The settling pond created at the foot of an oak wood
05. Pontoons on the iris pond
06. Teenagers watch wildlife from the hide
07. The new reed bed at the heart of a coppice landscape
08. Measuring the ecological dynamic

01 景观与大型基地 Landscape and site

09. 收集流水的结构体同时也引导着游客前往公园
10、11. 公园中的平台面对着高密度的城市
12. 亲水阶梯
13. 浅滩过道
14. 为景观带来新风格，集水沟过桥处理
15. 雨水调控设置是预铸的水泥结构体

09. The rainwater harvesting ditches guide the walk through the park
10-11. The park terrace, facing the new dense town centre
12. Steps down to the water
13. Stepping stones
14. A new look to the landscape, and the water-harvesting ditch
15. The structures for channelling rainwater run-off are in prefabricated concrete

01 景观与大型基地 Landscape and site

露天运动游憩区
Sports and Leisure Park

COULON LEBLANC & ASSOCIÉS

地点：法国托尔西
完工日期：2013
面积：12 ha
业主：法兰西岛（大巴黎）区政府
照片版权：Jacques Coulon - Linda Leblanc

Location: Torcy, France
Completion date: 2013
Area: 12 ha
Client: Ile-de-France Regional Council
Photo credits: Jacques Coulon - Linda Leblanc

01. 湖畔浴场平面配置图
02. 水滑梯、浮岛和沙滩平台
03. 面向正南的白杨树提供了舒适的遮阴
04. 喷水游戏与大水池

01. Plan of the bathing area
02. Water chutes, floating islands and beach platforms
03. Facing south, in the shade of the white poplars
04. Fountains and the large lake

托尔西是城市化密度极高的大巴黎地区的一个特殊地域，它不仅是一个户外游乐场地，更是一个优美的大地景观。托尔西丘陵脚下自然生成的大型湖泊是马恩河河床几度移动所留下的地理痕迹，为人们带来特殊的场所感受：风、阳光、旷阔、水影、白杨林里颤动的树叶、湖面飞鹅的啼声、滩边啄沙的鹡鸰。从巨大到微小，在在呈现出基地的质感，撩拨着人们的感官与情绪。

此游憩区自从1980年开放之后，原本简约的基地随着逐次的整治规划而丰富起来，强化出场所的特殊氛围。2012年展开的新计划包括游戏设施、水滑梯、水上活动、漂浮码头、固定浮桥、木造观察站和桥道等建设，此计划提供了展现马恩河主河床景观的崭新机会。一条柔性交通小径将近处的河岸和掩藏在远处茂密林地后的河岸联结了起来。

向南的大型草坪是日光浴场的所在地，其曲线边缘新种植了树干高耸的槐树并且设置了木质长凳，以提供遮阴休憩。当初为了建设基地而开采砂石的灰尘与噪音早已被人遗忘，如今每到夏日晴朗的周末，便有上万人前来享用这个景观游憩区。

Torcy, this exceptional breathing space in the dense urban environment of Ile-de-France, is more than a leisure park, it is a landscape. Here you can experience the joy of getting back to nature, thanks to the large lake naturally installed at the foot of the Torcy hillsides, in the geographical path of the fluctuating bed of the Marne: the wind, the sun, the wide open spaces, the reflections in the water, the leaves that rustle in the poplar trees, the cry of the geese flying over the lake, the wagtails that graze on the sand. From the smallest feature to the largest, everything plays a part in this environment, and everything provokes emotions.

Since its opening in 1980, the simplicity of the site has been enhanced by waves of improvement that have given it yet more atmosphere. In 2012, the addition of playful features including water chutes, fountains, floating landing stages and fixed pontoons, a lookout tower and a wooden footbridge, were a chance to create a new setting for the great landscape of the Marne flood plain. Soft transport routes link the banks that are so near but were until recently masked by wooded masses.

The curve of the sun-bathing area – a large, south-facing lawn – is shaded by new plantings of high-stemmed Japanese pagoda trees, accompanied by large hardwood benches. Today, everyone has forgotten the dust and the noise of the gravel pit that led to the creation of the site, and 10,000 visitors descend here on sunny summer weekends.

05. 沙滩、长凳与浮桥
06. 喷水游戏与湖畔浴场
07. 沙滩和救生员观望台
08. 沙滩平台

05. The beach, benches and pontoons
06. Fountains and the bathing lake
07. The beach and the lifeguard tower
08. A beach platform

01 景观与大型基地 Landscape and site

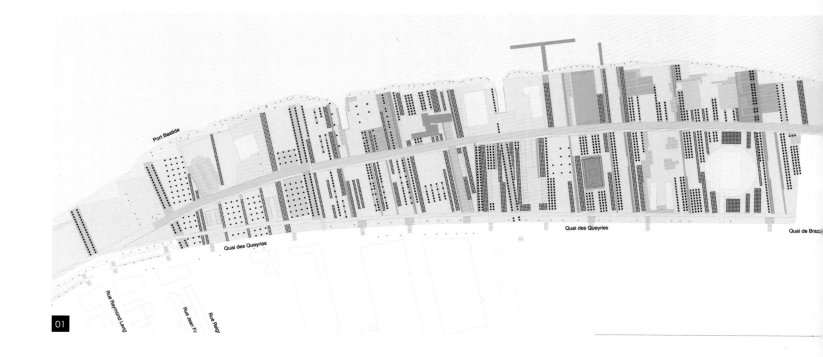

01

安洁丽克公园 *Angéliques park*

MICHEL DESVIGNE PAYSAGISTE

地点：法国波尔多
完工日期：2010-2017
面积：75 ha（第一阶段10 ha）
业主：法国波尔多市政府
照片版权：Michel Desvigne Paysagiste (n°04, 05), Guillaume Leuregans (n°06-08)

Location: Bordeaux, France
Completion date: 2010-2017
Area: 75 ha (10 ha – 1st phase)
Client: Bordeaux City Council
Photo credits: Michel Desvigne Paysagiste (n°04, 05), Guillaume Leuregans (n°06-08)

01. 科里斯堤岸段落：第一阶段竣工范围
02、03. 研究模型
04. 加龙河右岸的重新整治行动
05. 此整治方案成为街区多个城市项目的定位点

01. Queyries sequence: the first phase to be completed
02-03. Study model
04. An operation to reclaim the right bank of the Garonne
05. A geographical anchoring for several new neighbourhood projects

加龙河的右岸正在经历一场转化，其港口和工业活动很久前就已经迁移，为波尔多市提供了建设一个重要户外生活空间的契机，一个位于市中心的公园与河左岸的历史建筑以及近期建设的花园式散步道隔河相望，将成为城市的主要公共空间之一。如同所有城市变迁一样，这个河岸的转化将持续几十年的时间。

项目管理者制定了一个非常务实的分期更新计划：每片空闲的土地都会在适当的时机快速地被绿化。此公园作为一种具有调节性的自然空间而伴随着城市变革，逐步为即将形成的新街区提供高质量的环境。安洁丽克公园成为项目第一个实施的空间，其中种植的灌木丛围合着若干林中空地，形成一个个公共场所。

成排的树木以不规则方式排列着，但其构成的不同形态直线都与河岸垂直，产生有如舞台的滑槽隔墙一般的作用。它们沿着与河流平行的散步道勾勒出森林的剪影，其密度、透明度和疏松程度都在不断变化。根据旧工业地块的朝向，这些"滑槽隔墙"以流畅的方式引导着视线的方向、组织着朝向河流的路线，同时也为未来的街区预先建立地理标志。

The right bank of the Garonne is undergoing a transformation. For a long time, port and industrial activities have been moving away, offering the opportunity to create a major link for Bordeaux: a huge park in the town centre, facing the historic facade of the left bank and its recent planted walks. As with any urban mutation, this transformation will take several decades.

The project envisages a very pragmatic process of progressive substitution: as opportunities arise, each vacant surface will immediately be planted. The park is a kind of intermediate natural environment that evolves hand-in-hand with urban change, progressively improving the surroundings in order to usher in the new neighbourhoods. The Angéliques park will be the first element to be completed. It is made up of copses marking out clearings that provide public spaces.

The rows of trees, planted irregularly but running perpendicular to the riverbank, function like the wings of a theatre: along the walks parallel to the river, they are forest silhouettes whose density, transparency and porosity vary. According to the arrangement of the old industrial parcels, these "wings" frame the views and organise the movement of people towards to the river with great fluidity, prefiguring the geographical anchoring of the future neighbourhood.

06. 此公园为波尔多城乡区域的居民提供了一个新的中央公共空间
07、08. 省事而有效率的方法：着重设计概念的严谨性和清晰度

06. A new central public space for the urban area
07-08. Economy of means: a design that is rigorous and easy to read

01 景观与大型基地 Landscape and site

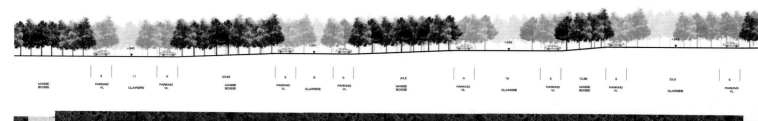

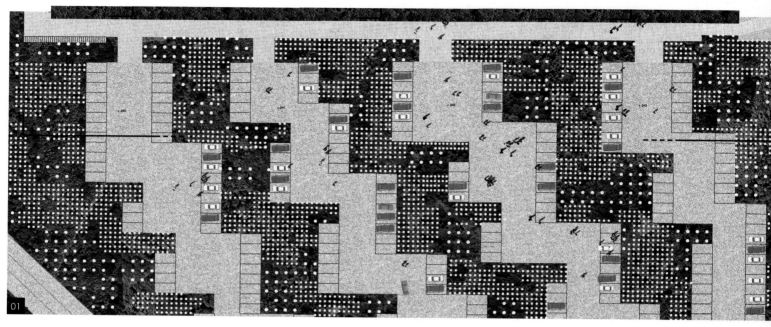

阿莱西亚博物馆公园
Alésia MuseoPark

MICHEL DESVIGNE PAYSAGISTE

地点：法国阿利兹-圣-莱纳
完工日期：2010
面积：3,7 ha
业主：黄金海岸省议会
合作设计师：BTUA Bernard Tschumi Urbanistes Architectes
照片版权：Michel Desvigne Paysagiste

Location: Alise-Sainte-Reine, France
Completion date: 2010
Area: 3.7 ha
Client: Côte d'Or General Council
Co-projet manager: BTUA Bernard Tschumi Urbanstes Architectes
Photo credits: Michel Desvigne Paysagiste

01. 以齿状林中空地作为空间的组织形态
02. 方案看似一座雕塑森林
03. 一系列向远方散开的林中空地
04. 可行驶汽车的草地，不设有任何看得见的结构物

01. Composition of staggered clearings
02. The project gives the appearance of a sculptured wood
03. A succession of clearings with open views
04. A grassy surface that can be driven over, with no visible structures

阿莱西亚博物馆公园表演中心的停车场整治方案犹如在这片充满历史的广阔大地上展开一场针灸术。重点并不是要把停车场隐藏起来，而是使它成为让人们认识这块土地的重要场所。

这片停车场成了一个容纳了多种不同场所的森林，停车空间便分布在几个尺度宽大、铺着草地的林中空地当中。在这里，预留给行人的疏散空间相当充足，不像大部分停车场一样，只在成排的汽车后面设置一条狭窄的通道。由于行人空间占有很大的重要性，因此司机在寻找车位时需要缓慢行驶。这个停车场确实满足了停车的功能需求，但它更像是一个乡间散步的场所。为了与周围的景观环境相对照，它以一系列"林木隔墙"和林中空地交错相间的形态呈现，在空地中人们可以欣赏到远处的风景。

这些新创的小林地是邻近大森林的微缩转化，壮观的风景也由此被内在化，两种尺度的景观自然地交织在一起。

The laying out of the car park around the Alésia MuséoPark interpretation centre can be seen as a kind of acupuncture taking place on this large historic site. The challenge was not to hide a car park but to make it an important place helping people to discover this area of land.

The car park is a wooded place integrating different kinds of environments, a wood sculptured around spacious grassy clearings. Here, there will really be space for pedestrians to move. Unlike many car parks, it will not have narrow paths running behind the rows of cars. In consequence, motorists will circulate slowly while looking for a parking place, because the space has been given over to pedestrians. Of course, this car park fulfils its function, but at the same time it evokes a walk in the countryside. Confronted by the landscape that frames it, it adopts a profile of "wooded corridors" and successive clearings, from which the views extend into the distance.

The wooded areas have been conceived as a miniature transposition of the neighbouring woods. The larger landscape is thus interiorised, and a relationship is naturally woven between the two different scales of landscape.

05. 这些新创的小林地是邻近大森林的微缩转置
06. 种植密度的"森林式"管理
07. 景观尺度超越了项目计划的实用性尺度

05. A transposition in miniature of the surrounding woods
06. Forest-style management of the plant densities
07. The landscape achieved goes beyond the utilitarian purpose of the project

01 景观与大型基地 Landscape and site

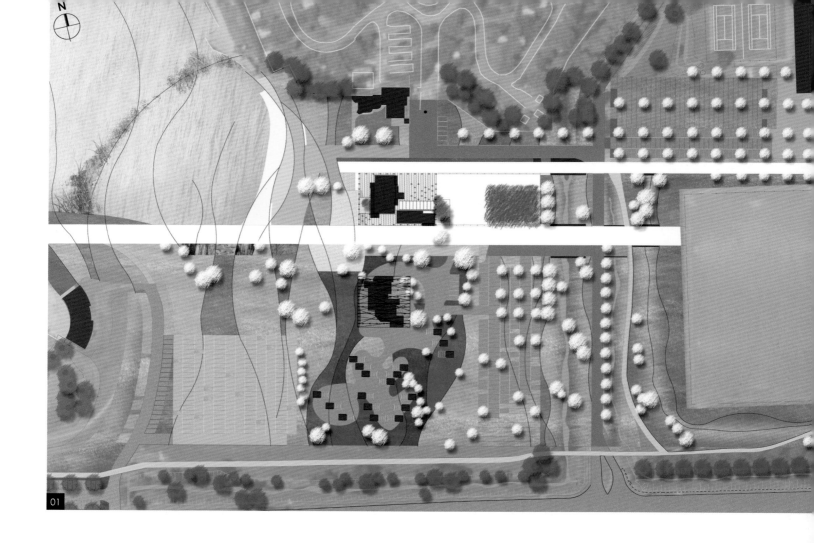

维祖尔湖游憩区
Vesoul Lake Leisure Park

DIGITALE PAYSAGE

地点：法国维祖尔
完工日期：2011
面积：5 ha
业主：维祖尔城乡区域市镇联合组织
合作设计师：Architectures Amiot Lombard
照片版权：Nicolas Waltefaugle (n°02, 04, 06, 08), Agnès Daval (n°03, 07), Nova Flore (n°05)

Location: Vesoul, France
Completion date: 2011
Area: 5 ha
Client: Communauté de Communes de l'Agglomération de Vesoul
Co-project manager: Architectures Amiot Lombard
Photo credits: Nicolas Waltefaugle (n°02, 04, 06, 08), Agnès Daval (n°03, 07), Nova Flore (n°05)

01. 竞赛方案的总体平面配置图
02. 可观赏湖泊的浮桥
03. 北边的"条板散步道"可直接通达湖畔

01. Competition master plan
02. Pontoon looking towards the lake
03. The north boardwalk with a view of the lake

维祖尔湖地区是30年前开发建造的，直到今天在维祖尔居民心目中，它仍然是一个具有代表性的重要场所。然而目前，它却呈现出陈旧的面貌，各项设施都已不堪重负。这个整治计划为湖区的散步场所带来新的前景，展现出周边的环境景观，并使交通流线组织更为清晰明了。

游乐区入口处的空间组织和流线安排非常简单而明晰，它强制人们把汽车停在上游处，并且在运动区和湖区之间重新建立了一条具有高安全性的柔性交通系统。两条东西向的主要行人散步道，即所谓的"条板散步道"，通过中央广场而互相联调，它们个别在基地的北方和南方连接了具有不同使用功能的场所。湖边的散步道终于找到与其相适的尺度。

在景观方面，方案必须在植物与城市、自然与娱乐之间寻找到和谐的平衡点，让每个元素都在其中各得其所：从翠鸟到周日的散步者，从蜻蜓到来自荷兰的观光客。为了善用这个场地的"潮湿"特性（粘土质土壤），设计师在地势变化上作了很多细微的规划，这些不同空间以"干地"（南侧和北侧的条板散步道、中央广场、中央停车场的前端）的标高作基准；用来吸收局部洪泛的草地，即所谓的岸滩，标高则较低一些；栽种植被的斜沟、水渠和堑沟则标高更低，并以芦苇过滤系统用来净化径流水质。

Developed 30 years ago, Vesoul lake has always been a symbolic place cherished by the inhabitants of Vesoul. It was suffering from an outdated image and facilities in need of renewal. The project has given the lakeside walk a new view, and simplified the circulation whilst making the most of the surrounding landscape.

The entrance to the leisure park is organised according to a composition and distribution that are simple and easy to read. Placing the parking near the entrance, it re-establishes a safe environment for walking and cycling between the sports facilities and the lake. Two wide east-west walks, "the boardwalks", linked by the central square, feed the different functions of the site to the north and south. The lakeside walk, a central feature of the site, finally has an approach worthy of it.

In terms of landscape, the project was about rediscovering a harmonious balance between plants and the urban environment, nature and leisure pursuits, where everything and everyone has its place: from the kingfisher to the Sunday walker, from the dragonfly to the Dutch tourist. To make the most of the "wetland" nature of the site (clay soil), the project pays special attention to the levelling, with a benchmark altimetric level for the "dry feet" areas (the north and south boardwalks, the central square, the first tier of the central car park), lower levels for the floodable meadows and the beach, and still lower levels for the swales, the rivulets and the planted ditches dedicated to the treatment of run-off water by a system of reed bed filtration.

04. 野餐场地
05. 花朵盛开的草原，2011 年6月
06. 南边的"条板散步道"
07. 南边浮桥的细部
08. 方案特别保留的一棵老树

04. Picnic area
05. The wildflower meadow in June 2011
06. The south boardwalk
07. Detail of the south pontoon
08. Tree heritage preserved by the project

01 景观与大型基地 Landscape and site

阿里·帕鲁萨公园
Alye Parusa Park
DVA PAYSAGES / OLIVIER DAMÉE

地点：俄罗斯沃罗涅日
完工日期：2011
面积：10 ha
业主：沃罗涅日市政府、沃罗涅日区域政府
合作设计师：Mégapark / Karina Lazareva, Marina Bailozian
照片版权：DVA Paysages / Olivier Damée

Location: Voronezh, Russia
Completion date: 2011
Area: 10 ha
Client: Voronezh City, Voronezh Region Council
Co-project manager: Mégapark / Karina Lazareva, Marina Bailozian
Photo credits: DVA Paysages / Olivier Damée

01. 开向城市的一扇窗
02. 松林氛围
03. 公园平面配置图

01. The window on the city
02. Pine forest atmosphere
03. Master plan of Alye Parusa Park

"阿里·帕鲁萨"公园的重整和扩建项目是在沃罗涅日市庆祝城市建成425周年的时机实现的。

公园主题的灵感来源于亚历山大·格林的小说"阿里·帕鲁萨（红帆船）"，该项目旨在重新审视传统俄罗斯休闲公园并创造一个新世界，让人们在其中享受到远离城市的新鲜感。方案设计特别保持并加强了这个犹如天然珍宝盒的基地的特色，尤其强调了面向沃罗涅日"海洋"的视野，这同时也是公园构成的关键元素。重现并改造这部俄罗斯文学作品中所描述的氛围则成为这个公园的设计主线。

改造后的公园向参观者展示了各种用途的空间。首先是以文化功能为主导的露天剧场，可用于露天电影的放映、表演或者音乐会。接下来是水边的休闲空间，包括餐馆、咖啡厅和散步道。最后是娱乐空间，让人们产生远离城市的感觉，林中空地为人们提供不同的游憩地（运动区、儿童游乐区……）。这个距离沃罗涅日市中心仅仅几分钟距离的公园，从其中心一片新建的水岸散发出独特的魅力。

The project for the restructuring and enlarging of Alye Parusa park was carried out for the 425th anniversary of the city of Voronezh.

On a theme inspired by the Alexander Grin novel *Alye Parusa* (*Crimson Sails*), the project aimed to revisit the tradition of Russian leisure parks and create a new world where citizens could escape the city and feel they were getting away from it all. While preserving and enhancing this natural scene, the project in particular improved the viewpoint over the "sea" of Voronezh, a key feature in the composition of the park. Recreating the atmosphere of the literary work was the guiding line of the design.

The layout of the park reveals spaces with various uses for visitors. First of all those with a cultural vocation, like the open-air theatre, which can be used for film projections, plays and concerts. Next spaces for relaxing beside the water, with a restaurant, a café, places to walk. And finally recreational spaces, creating the feeling of being far from the city, with clearings for different activities (sports grounds, children's playgrounds…). A new beach provides a unique attraction at the heart of the park only a few minutes from Voronezh city centre.

04. 露天剧场的座椅细部
05. 露天剧场
06. 沃罗涅日的沙滩与海洋
07. 公园的新沙滩
08. 从松林望向海洋

04. Seating in the open-air theatre
05. The open-air theatre
06. The beach and the sea of Voronezh
07. The new beach of Alye Parusa Park
08. A window on the sea

01 景观与大型基地 Landscape and site

A Path through the Pine Forest

松林下的步道

AGENCE HORIZONS / JÉRÔME MAZAS

地点：法国卡尔维
完工日期：2012
面积：14 000 m²
业主：卡尔维镇政府
合作设计师：Pozzo di Borgo
照片版权：Nicolas Faure & Jérôme Mazas

Location: Calvi, France
Completion date: 2012
Area: 14 000 m²
Client: Calvi Town Council
Co-project manager: Pozzo di Borgo
Photo credits: Nicolas Faure & Jérôme Mazas

01. 日暮下的步道
02. 步道因应环境条件而加宽
03. 沿着水岸延伸的木板步道

01. A walk at the end of the day
02. The path widens
03. The wooden walkway beside the sea

卡尔维海湾是一个神奇的地方，每天变化的光线、清晨的薄雾、咸涩的气味和松树的芳香微妙地混合在一起。市政府要求方案保留位于一片巨大松林旁、第一个小沙丘边缘的铁路通道，同时要求为连接这片松林和市中心的散步道建立舒适的环境。在这个被登记保护的地块中，市政府很长时间以来通过各种方法来维护这片环境，例如：种植植被、进行保护措施以及设置被细木围栏所引导的步行道。

景观师杰侯姆·玛扎斯利用在沙丘边缘和松树林中出现的植被，精心设计了一个简洁的方案：这是一条飘浮在沙丘上的木板小路，随着沙丘的地势而起伏并让风带来的细沙任意穿过；它或者绕过高大的松树，或者将它纳入到它的路线中。沙滩的一侧种植了大量的植物，与现有植物群连成一片。这个设计将沙子引向沙滩，也令沙滩更加舒适，同时，在当地苗圃的协助下，以不同植物的选择为基地带来生物多样性。在一些可以凝赏大海的地点，步道时而被适度地加宽而形成休息平台，让人眺望远处的卡尔维海湾、城堡和巨岩。

如今，这个方案在美丽和壮观的景色面前逐渐消失了：木头的使用正是为了使方案以简朴的方式面对大自然，新种植的作物成为原有植物的延续，而使用者则享受着大海的壮丽风景。

The bay of Calvi is a magical place, a subtle mixture of constantly changing light, morning mists, briny smells and the perfume of the pines. The project commissioned by the municipality consisted of protecting the sleepers of the railway that skirts the first dune, on the edge of a large pine forest. It also involved ensuring there was a comfortable walk between this pine forest and the town centre. The municipality has looked after this registered site for a long time, putting in plantations, protective measures and pathways accompanied by windbreak fences.

From the vegetation present on the dune ridge and in the pine forest, Jérôme Mazas drew up a simple project: a wooden path floating on the dune that follows its contours and allowed the windblown sand to go through, a wooden path that avoids the large pines or includes them in its design, and a sizeable planting of vegetation on the beach side that continues the plant formations already in place. This approach allows one to catch the sand near the beach and to buffer it too. It also allows for increasing the diversity through a varied choice of plants, achieved with the help of the local nurseries. The pathway has been widened in places where people often stop to contemplate the sea to form rest beaches with the bay of Calvi, its citadel and its rock masses in the far distance.

Today, the project is subsumed by the force and the beauty of the site: the use of wood allows it to remain discreet, the planted vegetation prolongs what already existed and walkers enjoy greater comfort and the sublime views of the sea.

04. 穿越松林的步道
05. 沿靠道路的步道
06. 通往"松林边缘"停车场和沙滩旁的柽柳
07. 步道局部加宽以提供休息和观赏的空间
08. 金属、细沙和木板
09. 如何建造铺板?
10. 透过空间的整治来保护沙丘与松林

04. The path through the pine forest
05. The path next to the railway
06. Entrance to the Orée des Pins car park and the tamarisk trees on the beach side
07. The path widens, providing an area for sitting and admiring the view
08. Metal, sand and wood
09. How to build a decking?
10. Protecting the dunes and the pine forest while laying out a path

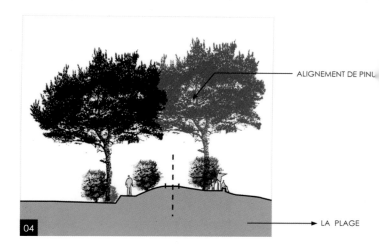

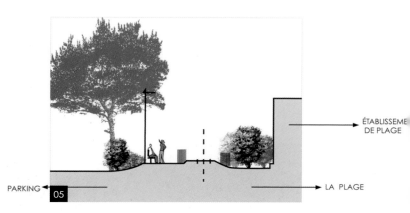

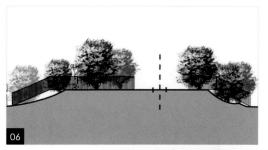

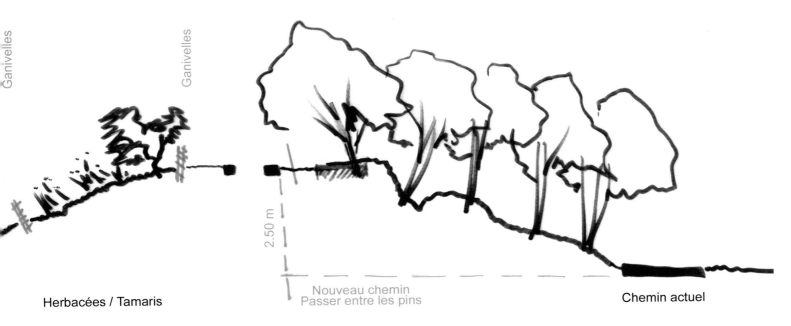

Herbacées / Tamaris   Nouveau chemin
                           Passer entre les pins   Chemin actuel

2.50 m

Ganivelles

01 景观与大型基地 Landscape and site

11. 松林和其生态环境，禾本植物位于前景
12. 如何避开这些松树
13. "松林边缘"停车场的植物栽种

11. The pine forest and its environment, with grasses in the foreground
12. How to avoid the pine trees?
13. Planting at the Orée des Pins car park

01 景观与大型基地 Landscape and site

01

马恩河岸
Banks of the Marne
FLORENCE MERCIER PAYSAGISTE

地点：法国托尔西
完工日期：2011（第一阶段）
面积：2.5 ha
业主：瓦勒莫布埃新城乡区域理事会
照片版权：Florence Mercier

Location: Torcy, France
Completion date: 2011 (1st phase)
Area: 2.5 ha
Client: SAN Val Maubuée
Photo credits: Florence Mercier

01. 总体平面配置图
02. 一个悬架在马恩河上方的平台和空地
03. 贡多瓦溪和马恩河汇流处的整治
04. 在河岸与森林之间的散步道

01. Overall plan
02. A clearing and a pontoon overlooking the Marne
03. Landscaping the confluence between the Gondoire and the Marne
04. A walk running between the banks and the woods

方案位于马恩河右岸,它的目的是在大区域尺度下,保持马恩河沿岸步道的连续性,并通过开放面向河流的新景观,来重新建立城市与河流的联系。

基地的大部分面积被品种单一的杨树林所占据,方案通过特殊的植物工程而巩固了被河流侵蚀的河岸,并通过促进新的生态系统的建立而丰富了生物的多样性。河岸步道和穿越道的设置,使得居民们得以重新接近水岸,在漫步中享受一系列引人入胜的景观。

步道与各种特殊的地形相得益彰,时而点缀着空地和浮桥,提供人们多种亲水的方式。斜沟花园通过建立不同形式的自然湿地、利用马恩河和贡多河的潮汐和地下含水层的压力变化,重新组合了新的河流景观。这个河流景观通过水体的诗意,寻找到了一个能将使用功能和生态丰富性结合在一起的全新诠释方式。

Situated on the right bank of the Marne, the project's objective is to ensure the continuity of the paths along the banks on the scale of the land area, and to re-establish the links between the city and the river by opening up new views towards the latter.

On a site whose planting was largely confined to a single species of poplar, the project will use plant species to stabilise the banks and enrich the ecological diversity of the site by encouraging new ecosystems to establish themselves. The laying out of the riverside path and crossings, which provides a more attractive setting for the walk, will allow the inhabitants to reclaim the banks.

Punctuated by clearings and pontoons, it offers several ways of experiencing the water, playing with the different conditions of the site. The swale garden composes new river views by offering a variety of natural wetland environments, making use of the flooding of the Marne, the fluctuations of the water table and the overflow of the Gondoire. This riverside landscape thus finds a new expression, bringing together uses and ecological richness, through the poetic shapes carved by the water.

05. 在河岸高处的木板铺地小径
06. 向马恩河敞开的林中空地
07. 一个与地势结合并伸出水面的浮桥散步道
08. 位于林中空地里的小广场连接着来自各方向的步道

05. A pathway in wooden decking on the crest of the riverbank
06. Clearings with views of the Marne
07. A pontoon-walk that hugs the land and ventures out into the water
08. Small squares articulate the paths at the heart of the clearings

07

08

67

01 景观与大型基地 Landscape and site

09. 穿越河岸与未来草沟花园的剖面图
10. 跨越贡多瓦溪的桥道
11. 穿越小径
12. 一个观赏景观的休息处
13. 面向马恩河的窗口

09. Cross-section of the banks and the future swale garden
10. A footbridge over the Gondoire
11. Cross path
12. A pause to enjoy the landscape
13. A window on the Marne

01 景观与大型基地 Landscape and site

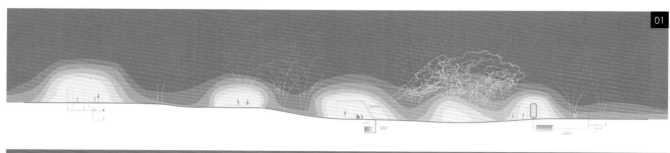
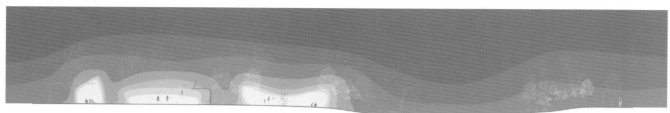
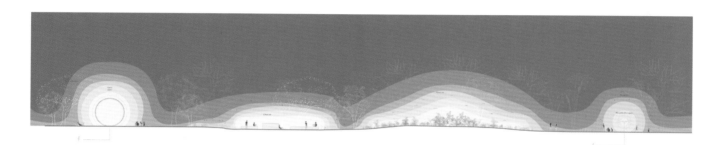

清翠园 – 相移公园
Jade Eco Park - Phase Shift Park

CATHERINE MOSBACH

地点：中国台湾台中
完工日期：2011-2015
面积：70 ha
业主：台中市政府
共同作者：Philippe Rahm
图片版权：Philippe Rahm (N°01, 02), Catherine Mosbach (n°04-06, 08, 09), Catherine Mosbach & Philippe Rahm (n°03, 07, 10, 11)

Location: Taichung, Taiwan
Completion date: 2011-2015
Area: 70 ha
Client: Taichung City Council
Co-author: Philippe Rahm
Image credits: Philippe Rahm (N°01, 02), Catherine Mosbach (n°04-06, 08, 09), Catherine Mosbach & Philippe Rahm (n°03, 07, 10, 11)

01. 气候路径剖面图：红色-温度，蓝色-湿气，灰色-污染
02. 初步研究
03. 竞赛阶段研究

01. Climatic path: red – heat, blue – humidity, grey – pollution
02. Preliminary design studies
03. Competition study

此项目通过公园的设置重新让居民得以享受户外生活，台中地区过热的自然气候在公园中获得缓和，使人们能够在较低温度、低湿度和低污染的户外环境里散步。公园空间结构的原则建立在因周围新建街区影响而产生的气候环境差异，使其空气更为凉爽、干燥和清净。一些气候分析图根据特别的大气参数而建立：分别展现温度、湿度与空气污染程度。这些分析图重叠之后得出多样的微气候，各种不同的微气候环境被使用于适当的活动与设施。

这些有关空气品质的分布研究还与不同土壤形式的分布研究相结合。公园中集水盆地所汇集的雨水与公园本身的关系决定了地膜覆盖方式以降缓水流速度，并使其尽可能在雨水落下后不远处便渗透到土地表层。土壤不同程度的渗透性与表面植物覆盖，决定了径流系数，并且依降雨程度而设定蓄水范围和临时淹浸的范围。水的地层研究与植物群系的地层研究也相互产生关联。这些参数使得在公园地形上塑造出的洪泛区会依据梅雨季的暴雨节律而充水，让居民见识到公园中70公顷面积的形变。

This project gives fresh air back to the inhabitants through a park where the excesses of Taichung's natural climate are softened: they will be able to enjoy outdoor walks in an atmosphere that is cooler, less humid and less polluted than the city around it. The structuring principle of the park is based on several layers of climate ambiances, echoing the new neighbourhoods created around it, which have also improved on the city's air quality. Climate charts have been drawn up for each specific atmospheric parameter: a map for heat, a map for the variations in air humidity and another for the amount of air pollution. These maps are superimposed, creating a variety of microclimates and numerous different ambiances, which have been used to plan uses and facilities adapted to each climate station for the best appropriation by the inhabitants.

The maps of air quality are matched with maps of the various ground treatments. The relationship of the park to the water that it harvests from its catchment areas is managed: membranes to slow down the water and filter it into the upper layers of the soil as closely as possible to the place where it falls. The different grades of porosity as well as the plant cover determine the run-off coefficients and direct the water to the retention beaches and the provisional submersion beaches according to the intensity of rainfall. The stratigraphy of the water is twinned with a stratigraphy of plant formations that is directly indexed to the run-off coefficients. In this way the topographical beds moisten according to the rainfall of the monsoon, offering the inhabitants a park that "morphs" over an area of 70 hectares.

04. 岛屿环境影响
05. 代表性生态环境研究取样
06. 初步设计研究
07. 一个"地形床"在雨季之外的使用

04. Insular influences
05. Transects of the environments represented
06. Studies before the project began
07. Appropriation of a "topographic bed" outside the monsoon period

通过对空气与地表条件的分析所识别出的空间特质，经过一系列气候装置的调和之后得以设置适当的功能活动。因此，在较为不舒适的环境建造了带有空调的封闭式建筑，而较为舒适的地点则设置具有休闲游乐性质的活动。

The park's facilities are organised according to the atmospheric data gathered by a catalogue of climate monitoring instruments. These measure the qualities identified by the atmospheric mapping, combined with those of the lithosphere. So the places with the most uncomfortable climate have been equipped with closed buildings with air conditioning while the recreation spaces are situated in the most comfortable atmospheric stations.

Rivers and coastline
farmland
Tropical savannas
Parks
Woodlands
Wastelands
Gateway park (Jade eco park)
Gaoemei wetland wildlife refuge

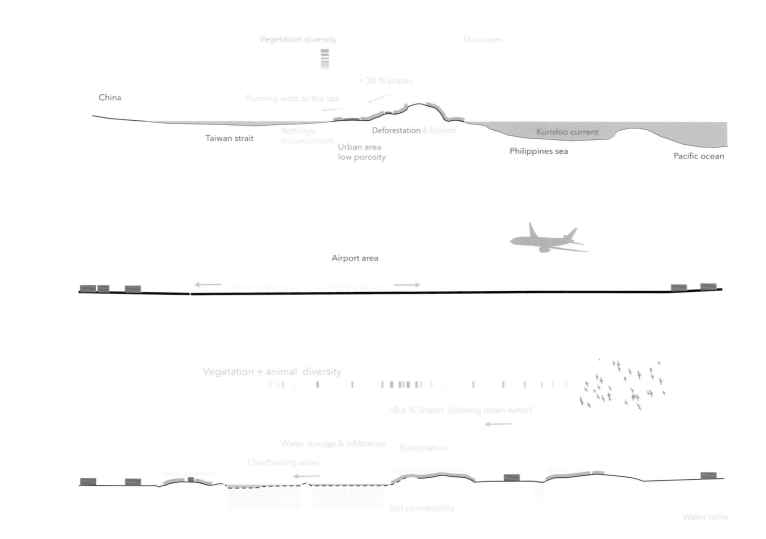

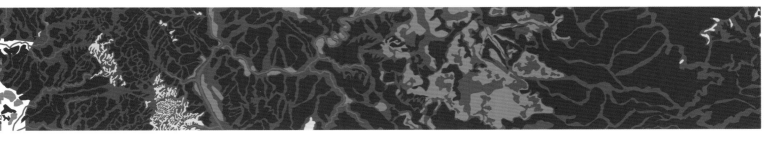

01 景观与大型基地 Landscape and site

	win	spr	sum	aut
Willow *Salix vinimalis*				
Butomus *Butomus umbellatus*	+	+	+	+
Deschampsia cespitosa *Deschampsia cespitosa*				
Menthe aquatica *Mentha aquqtica*				

08

Willow
Salix vinimalis

Butomus
Butomus umbellatus

Deschampsia cespitosa

Menthe aquatica
Mentha aquqtica

	gradient	x1	x2	x3	x4	inverse
material						
pattern						
membrane						
porosities						

09

| no-skid | interstices *cut grass* | planted area *grassy* | shrubs | furnitures *potelet* |

10

08. 种子传播
09. 地面材质渗透度
10. 竞赛阶段透视图：城市荒原
11. 竞赛阶段透视图：林下过道、多用途平原、设置配备的林中空地、自然与科技园地

08. Seed sowing
09. Porosity gradient
10. Competition visuals of the urban heathlands
11. Underwood paths, polyvalent plains, landscaped clearings, nature and technology

01 景观与大型基地 Landscape and site

朗斯罗浮宫公园
Louvre-Lens Museum Park

CATHERINE MOSBACH

地点：法国朗斯
完工日期：2005-2013
面积：25 ha
业主：北加莱海峡地区政府
共同作者：Sanaa 博物馆建筑师、Imrey Culbert 博物馆场景设计师
图片版权：Hisao Suzuki (n°01, 02), Iwan Baan (n°03), Catherine Mosbach (n°04-09, 11, 12), Catherine Mosbach & Sanaa (n°10, 13)

Location: Lens, France
Completion date: 2005-2013
Area: 25 ha
Client: Nord-Pas-de-Calais Regional Council
Co-author: Sanaa architecte, Imrey Culbert architecte museographer
Image credits: Hisao Suzuki (n°01, 02), Iwan Baan (n°03), Catherine Mosbach (n°04-09, 11, 12), Catherine Mosbach & Sanaa (n°10, 13)

01. 从大广场望向建筑北面前庭
02. 从旧矿井区望向博物馆休息室
03. 博物馆开幕时的鸟瞰全景
01. View of the grand esplanade towards the north forecourt
02. View from the wells towards the foyer
03. Aerial view at the opening

造访者从与洛桑戈埃勒丘陵的四方视野一样遥远的地方起步，沿着昔日骑士走过的金合欢林荫道前来。于入口处，行进的节奏在边界的明暗交替与林中空地的眩目晕然之间摇摆。随着林荫道的前进，花园与游客服务设施一一展现。在优先路径(或说是"快速"路径)的旁侧，一束小径邀请人们随意漫步(或说是"缓调"漫步)，观赏以事件平台形式所呈现的花园。

边缘地带的草原群系界定出林中空地公园的尺度：从东到西的高草原，出现一行行的低矮草地作为穿越小径；靠近街区有微型花园；靠近入口大厅则为草坪沙发和渗水圆环，展现出新生苗圃。

此景观方案仿佛一个存放材料的仓库，通过内外的交缠而展现轮廓。此材料库既与参观者的游览过程产生对话，也任由时间、水、植物对其发生作用，创造出同时是景观的也是人文的作品。此项目既非公园、亦非城市边缘的森林，而是一个具有先创性的博物馆园地。

From as far as the four horizons of the hills of Loos-en-Gohelle, visitors take the old bridleways covered with acacias. At the threshold, the rhythms oscillate between the dappled light of the borders and the dazzle of the clearings. Travelling through the bridleways, the gardens unfold as well as the facilities for visitors. Beside the priority walkways – the "fast-track" ones – an array of paths invites one to wander freely – the "slow-track" – from gardens to stages for hosting events.

The size of the park clearing is suggested by meadow-like formations flanked by borders: fields of high grasses from east to west, furrowed by paths of mowed grass; miniature gardens near the residential neighbourhoods; grass-covered sofas and haloes of filtered water nearer to the entrance hall; a procession of young plants everywhere in the underwood.

The contours of the project mingle exterior with interior using stockpiled materials. They welcome the changes wrought by the feet of the public and by time, by water, by plants, producing artworks created by the landscape and by Man. Neither public park nor peri-urban forest, this is a pioneering museum-park.

04. 景观产物犹如文化遗产资源
05. 不同层面的公园功能计划剖面图
06. 抵达博物馆的通道类型，2012 年
07. 从土壤动力到植物活力
08. 从餐厅露天平台望向博物馆东侧前庭

04. Landscape products as legacy
05. Programming section of the different strata of the park
06. Accessibility typology 2012
07. From soil dynamics to plants
08. The east forecourt from the restaurant terrace

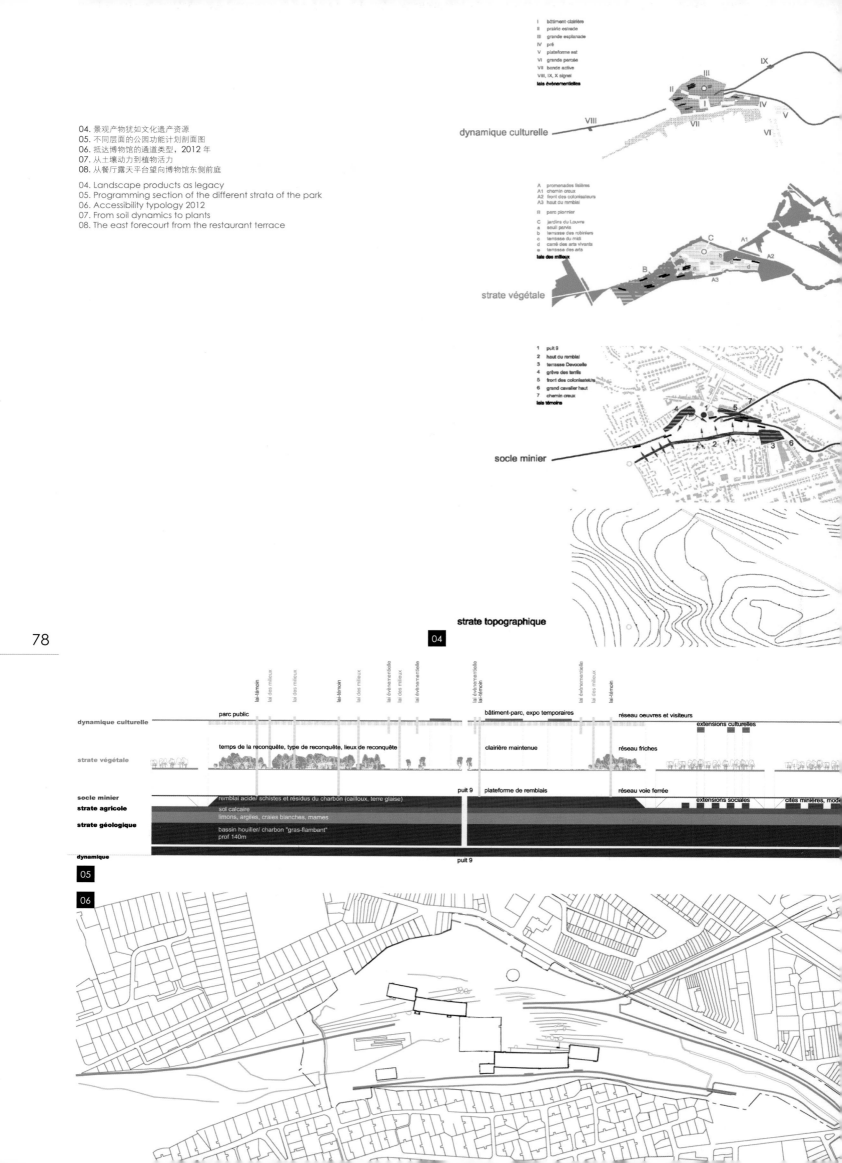

01 景观与大型基地 Landscape and site

09. 与先锋树林相邻的蓄水池
10. 艺术平台
11. 方案模型：昔日骑士走过的林荫道研究
12. 方案模型："草坪沙发"休息空间研究
13. 大草原

09. Reservoir on the edge of the pioneer wood
10. Arts terrace
11. Study of the brideways. Model detail design
12. Study of the resting places called "grass sofas". Model detail design
13. The meadow

01 景观与大型基地 Landscape and site

巴尔河生态公园
Barre Ecological Park
MUTABILIS

地点：法国安格雷
完工日期：2007
面积：15 ha
业主：巴约讷-安格雷-比亚里茨城乡区域联合组织
照片版权：Mutabilis

Location: Anglet, France
Completion date: 2007
Area: 15 ha
Client: Communauté d'Agglomération Bayonne Anglet Biarritz
Photo credits: Mutabilis

01-03. 多样化的路径和观景点带领人们认识基地的各样特色
04. 苗圃的设置：强化对生态环境的修复
01-03. There are a host of paths and viewpoints on the specificities of the site
04. Planting a nursery: making the most of the restoration of habitats

这个地处于阿杜尔河口、占地15公顷的生态公园位于沙丘背后，避免了水沫的冲击。公园最引人瞩目的是它的两个湖，一个是由阿杜尔潮汐形成的咸水湖，另一个则是由地下泉水形成的淡水湖。

公园中有四种被保护的珍稀植物，并特别具有反差明显的生态环境，从野生动物学或者地区植物学角度来看都具有很高的价值。这些都使得巴约讷－安格雷－比亚里茨城乡区域联合组织决定将此地改造成为一个生态公园。如果说以生态的方式修复自然环境是这次规划的中心思想，那么赋予活力、对外联系和改善开发等问题则是密不可分的课题，并且需要一个特别能够回应方案条件的对策。

Situated at the mouth of the Adour, this 15-hectare ecological park is behind the dunes, sheltered from the sea spray. It is remarkable for the presence of two lakes, one of which contains the brackish water of the Adour tides while the other is fed by freshwater springs.

The presence of four protected species of plants and above all of contrasting environments, interesting from a flora or fauna point of view, has led the Bayonne Anglet Biarritz joint metropolitan area (CABAB) to highlight and enhance this site by planning an ecological park. If the restoration and rehabilitation of natural environments is at the heart of the initiative, questions of activities, communication and enhancement are dissociable from it and call for solutions that are specifically adapted to the conditions of the project.

05. 长廊，作为入口广场和公园中心之间的过渡空间
06. 各种不同的小径，展现基地的多样尺度
07. 栈桥码头和淡水湖
08. 鼠尾草叶岩蔷薇形成的荒原

05. The gallery walk, a transition between the forecourt and the heart of the park
06. Varied paths show the site in all its dimensions
07. The pier and the freshwater lake
08. Moorland colonised by the sage *cistus salviifolius*

为了保持景观的完整性，生态公园避免安置任何用于教学的展示，甚至没有信息指示牌，让人们首先用触摸和感觉来认知场所。这种想法的深层意义便是极少主义，推崇简单的作为。为了在一个如此微妙和脆弱的基地上容纳大量游客，方案将主要的步行栈道与自然地面分离，在引导游客体前进的同时也使他们认识基地的不同生态环境和特殊性。其目的是推动游客用心体会、观察、善加理解这个脆弱的生态环境。公园里设置的设施试图提供"展现-隐藏"的景观（树林屏障、铜质瞄准器、远眺台、鸟类欢察站……）、充满趣味的探奇之旅（瞄准器、公园游览的GPS导航工具）或者实验性经验（接待花园）。

To respect the landscape integrity of the site, the ecological park has been designed without educational scenography or information panels, preferring to put the visitor directly into contact with the tangible, with its effect on the senses. The philosophy thus developed is minimalist and extols the simple gesture. For a site this delicate and fragile to be able to cope with a large number of visitors, the main paths have been raised up from the natural ground, channelling the visitors while allowing them to take in the different environments and unique features of this site. The challenge is to push the visitors into paying attention, observing and better understanding this fragile environment. The structures put to work in the park play a game of "show-hide" (wooden screens, copper sights, a lookout tower, a bird hide...), of discovery through play (the sights, a GPS piloting tool for visiting the park), and of the experimental (the welcome gardens at the entrance).

85

01 景观与大型基地 Landscape and site

structure fer à béton cintré → paroi

Rubus ulmifolius

Ronces à feuilles d'orme.

2,00

1,4

6 m.

09. 初步设计的研究图案
10. 公园的观景台，向广阔的景观开展
11、12. 应用在整个参观路径的原则：展现与隐藏
13. 鸟禽沙滩观察站：在隐秘处观望

09. Pre-project stage research sketch
10. Belvedere overlooking the park and the wider landscape
11-12. Hide observation principle developed along the park itinerary
13. Bird-watching hide: seeing without being seen

01 景观与大型基地 Landscape and site

湿地公园-蒙多德汉航空中心
Wetland Park – Montaudran Aerospace

OLM / PHILIPPE COIGNET

地点：法国图卢兹
完工日期：2010-2020
面积：50 ha
业主：大图卢兹地区城市联合组织
合作设计师：Seura / David Mangin Architecte（设计总负责），Arcadis
图片版权：OLM

Location: Toulouse, France
Completion date: 2010-2020
Area: 50 ha
Client: Communauté Urbaine du Grand Toulouse
Co-projet manager: Seura / David Mangin Architecte (project representive), Arcadis
Image credits: OLM

01. 整体平面配置图
02. 鸟瞰全景
01. Master plan
02. Aerial view

蒙多德汉航空中心在1917年和1933年之间是很多著名的飞行员，例如圣-艾修伯里、纪尧梅、迈尔莫兹等创造辉煌成绩的场所。如今，这个场地将被改造成一个新社区，将于2020年陆续完成研究机构、住宅、服务设施、公共设备和公共空间。

方案围绕着保留下来的飞机跑道交错组织建筑功能。这是一个宏伟壮观的水平空间，长达1700米，连接了所有步行流线，并成为一个组织运动和文化性集会游行的场所。与跑道垂直相交的三个园区，住宅园区、运动园区和湿地园区，在南部的子午运河和昂格里校园，以及东北部蒙多德汉社区和马尔拜尔社区之间建立了连续的公共绿化空间。这三个尺度与功能各异的园区同时承担着雨水的调节功能，根据所收集雨水的分量和质量、地面的渗透性、空间的使用功能以及植物的吸收量等因素，将整个基地规划成几个汇水盆地。

Between 1917 and 1933 the old aerodrome of Montaudran was the site of the exploits of great aviators such as Saint-Exupéry, Guillaumet and Mermoz. The site is today being transformed into a new neighbourhood where research programmes, housing, services, facilities and public spaces are due to appear by 2020.

The project organises the built functions in a staggered format around the preserved runway, a horizontal monument 1,700 meters long, which serves as a main thoroughfare for pedestrians and becomes the place for sports and cultural events. Perpendicular to the runway, three parks – one residential, one for sports and one a wetland environment – create a continuity of planted public spaces between the Canal du Midi and the Rangueil campus in the south, and the neighbourhoods of Montaudran and Malpère in the north-east. These three parks, with distinct scales and vocations, fulfil a role in stormwater management. The plan is to structure the site into several detention basins according to the quality and quantity of the water harvested the permeability of the soil, the project's uses, and the degree of absorption by plants.

03. 湿地公园
04. 湿地公园的地形、水文和植被
05. 湿地公园规划原则平面图与剖面图
06. 涨水原理

03. Wetland Park
04. Topography, hydrology and planting of the wetland park
05. Plan and cross-section of the principles of the wetland park
06. Principle for rising water level

南部的湿地园区在有限的范围内收集三种类型的雨水。它的作用是让雨水逐渐而缓慢地渗入泥灰黏土质的土壤之中，来调节30年一遇的洪水。同时，它的目的也在于在一个开放给社区中研究人员、学生和居民使用的公共空间中，发展湿地的生态环境。这个湿地园区由层层渐次的条状花园组成，使水在其间借助重力作用而自然流动。这些通过浮桥系统和柔性交通连接了克雷蒙·阿黛尔中心和技术研究所的条状花园，有利于亲水性植物对水质的净化，并加强了位于几米之外、艾尔河沿岸的现有植物群的阵容。

The wetland park in the south will concentrate three water typologies in a confined area. Its role is to regulate the 30-year flood while ensuring a slow and progressive percolation of the water in a mud-marl soil. The aim is to develop a wetland ecology on a real public space open to researchers, students and local inhabitants. The park unfolds in strips through which water circulates gravitationally, cleaned by hydrophilic plants, which strengthen the existing flora. Running several hundred metres along the River Hers, it links the Clément Ader Space and the Institution of Technological Research through a system of pontoons and soft circulations.

01 景观与大型基地 Landscape and site

02

Public Parks and Gardens

公 园 与 公 共 性 花 园

Galbert Garden
加 尔 贝 尔 花 园
AGENCE APS

地点：法国安纳西
完工日期：2007
面积：6 000 m²
业主：安纳西市政府
图片版权：Pierre Vallet (n°01, 04), Agence APS (n°02, 03, 05-11)

Location: Annecy, France
Completion date: 2007
Area: 6 000 m²
Client: Annecy City Council
Image credits: Pierre Vallet (n°01, 04), Agence APS (n°02, 03, 05-11)

01. 花园向高山与湖泊敞开
02. 观景平台、纪念水池和岩石花园
03. 中央大草坪是整合花园的组织性空间

01. The garden has a view of the mountains overlooking the lake
02. The belvedere, the memorial lake and the rock garden
03. The wide central lawn organises the garden

加尔贝尔协议开发区位于湖滨附近一个新住宅街区的中心，此项目为其勾勒出了主要的公共空间。此花园是一个人人可享用的绿色休闲空间，它同时也希望与城市的其他邻里性公共空间建立起东西向的联系：APS事务所设计了一条供行人和自行车使用的天桥，犹如一杠连接符号，以便跨越邻近基地的铁道，与布洛尼林荫道相接而通往湖边。

花园设计的当代手法旨在与周边环境建立密切的关联：材料的选取、符合周边自然环境条件的植被、有利于空间使用的氛围塑造以及视野的开拓都是被关注的焦点。此花园虽然不具有阿尔卑斯山区典型花园的外观，却以空间材质与组织表达出其独特而强烈的高山特性。

At the heart of a new residential neighbourhood not far from the banks of the lake, the project gives a foretaste of the main public space of the Galbert ZAC (comprehensive development zone). A place of relaxation and greenery open to all, the garden also sees itself as an east-west link with the other nearby public spaces of the city: the agency APS has proposed a pedestrian and cyclist footbridge as a true link and clear evidence of urban planning, which will span the railway that borders the site and connect with Brogny Avenue to provide access to the lake.

The garden's contemporary design tries to weave perceptible links with its environment: the choice of materials, the plant palette adapted to the conditions of the environment, the quality of the ambiances created for users and the visual opening are successful examples. Without giving it the forced look of an alpine garden, its materials and its story take note of this particular identity, which ties in strongly with its mountain geography.

04. 此景让人联想起山中湿地的生态环境
05. 一个与环境背景融合的方案
06. 游戏场巧妙地融入花园之中
04. Ecological evocation of wetland mountain environments
05. A project situated in its context
06. The children's playground fits neatly into the garden

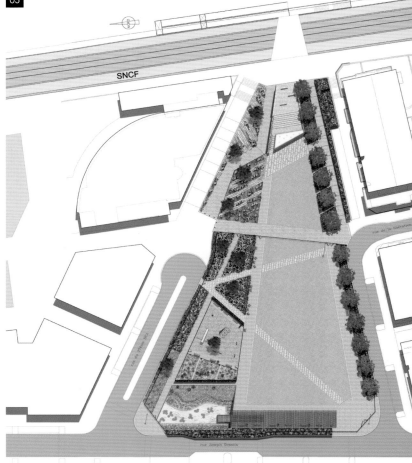

这块占地6000平方米的三角形基地，其形态和朝向都在很大程度上引导了设计构思与场景配置的实现。位于花园中心且具有整合作用的大草坪把形成基地架构的各种建造体、场所和主题联系在一起：一条提供遮阴的棚廊、数条配置有座椅和长凳的散步道、一个水园、一个周围环绕着多年生植物的儿童园、一道种植欧洲赤松的边界、一个杜鹃花园、一个由绣球花和苹果树组成的长向边缘带、一个由一层层宽厚石灰岩组成的岩石花园以及带有水墙的纪念碑；纪念碑的水池紧邻着位于花园底端的观景平台。

The original triangular geography and aspect of the 6,000 m² site have strongly guided the design and layout of the project. At the centre of the garden, the unifying space of the main lawn feeds the different openings, places and themes that structure the site: the gallery of the pergola-canopy, the lateral pedestrian walks with their chairs and benches, the water garden, the children's garden and the perennial garden, the border planted with sylvester pines, the rhododendron garden, the longitudinal margin of hortensias and flowering apple trees, the rock garden with its thick limestone strata, and the memorial, with its engraved vertical fountain and its pool next to the belvedere, at the far eastern end of the garden.

02 公园与公共性花园 Public parks and gardens

07. 岩石花园紧邻着中央草坪
08. 前阿尔卑斯山地带的景观联想
09. 岩石花园的材质搭配
10. 上萨瓦省法属北非纪念喷水池
11. 从观景平台高处望向中央草坪

07. The rock garden where it meets the central lawn
08. An evocation of the pre-alpine environment
09. Contrasting materials in the rock garden
10. The memorial fountain for soldiers from Haute-Savoie who died in action in North Africa
11. View from the highest point of the belvedere

02 公园与公共性花园 Public parks and gardens

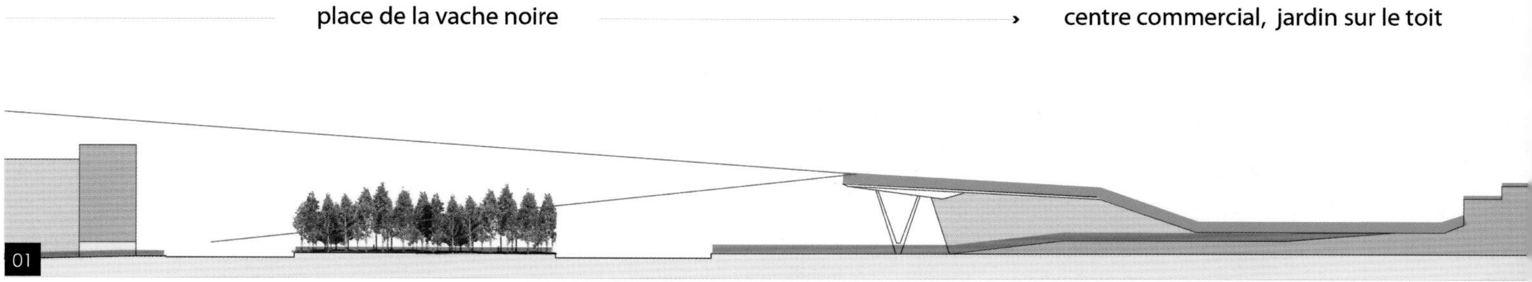

place de la vache noire → centre commercial, jardin sur le toit

01

Vache Noire Roof Garden

黑牛商业中心屋顶花园

AGENCE TER

地点：法国阿尔卡伊
完工日期：2007
面积：13 000 m²
业主：Multi Vest 3
合作设计师：Groupe 6 Architectes
照片版权：Agence TER / Yves Marchand & Romain Meffre

Location: Arcueil, France
Completion date: 2007
Area: 13 000 m²
Client: Multi Vest 3
Co-project manager: Groupe 6 Architectes
Photo credits: Agence TER / Yves Marchand & Romain Meffre

01. 设计构思原则剖面图
02. 商业中心的出风口成为屋顶花园动态十足的雕塑品
03. 在黑牛广场之上的观景平台
04. 研究模型

01. Main section
02. The air ducts of the shopping centre form dynamic sculptures
03. The terrace, a viewpoint over Vache Noire Square
04. Study model

阿尔卡伊是巴黎南部一个非常城市化的市镇，岱合事务所把黑牛商业中心的屋顶平台改造成一个公共花园以缓解这个地区匮乏绿色空间的现状。他们在这艘经济船舰的浮桥上设置了与购物活动形成对比、提供人们休闲、自由表达和景观惊喜的空间。

为了安排连接花园的出入口并且安置屋顶复杂的结构，景观师把屋顶折叠成起伏变化的表面。中心铺满草皮的部分是具有多种用途的"伸展台"。北面的草坡是"船尾"，作为相邻住宅的泻水坡。南侧的平台则是"船首"，它们构成的倾斜平面为人们提供了一片高出街道17米的观景草场。

技术设备也成为花园内的壮观风景：两个平铺的玻璃盒子既作为城市小品，勾划出散步道的边缘界限，同时也是采光井，为商业中心带来自然光线；夜晚则相反，商业中心的人工照明从玻璃盒子中透射出来装点花园。通风设备安置在7.5米高的金属网架中，它们被转化成为雕塑作品，或者从上到下遍布绿色植物，或者被丝网印刷玻璃包裹起来，从内部发出光芒。这些巨大的萤火虫在这片优雅简约的花园中营造出令人惊叹的感官效果。

In Arcueil, a very urban suburb south of Paris, Agence Ter has transformed the rooftop of the Vache Noire shopping centre into a public garden to compensate for the lack of green spaces. On the deck of this vessel of the economy they have installed a space for relaxation, for free expression and for landscape surprises that contrast with the shopping activities.

To facilitate the various access routes to and from the garden and to integrate the complex framework of the roof, the landscape architects folded the surface of the roof to obtain a dynamic topography. In the centre is a grassed-over surface with multiple uses: the "podium". A grassy slope in the north, the "stern", serves as a sloping lawn for the adjacent buildings. A balcony in the south, the "prow", is comprised of an inclined surface planted as a meadow offering a lookout point 17 metres above the street.

The technical elements have become the garden's most spectacular features: two flat skylights serve at the same time as pathways, urban furniture and atriums for the shopping centre, while at night the centre's artificial light illuminates the garden. The ventilation shafts, housed in metallic armatures 7.5 metres high, are transformed into sculptures that are half planted and half clad in screen-printed glass, illuminated from within. These giant glow-worms create a surprising sensuality for this simple and elegant site.

05. 高处平台：多用途的广大草坪空间
06. 玻璃天棚不仅是商业中心的采光井，也是屋顶公园的照明设施
07. 公园的活动与商业中心的人潮节奏相互呼应

05. The podium: a large grassy surface open to multiple uses
06. The glass roofs: light shafts for the shopping centre and lighting for the park
07. The park reacts to the frequentation of the shopping centre

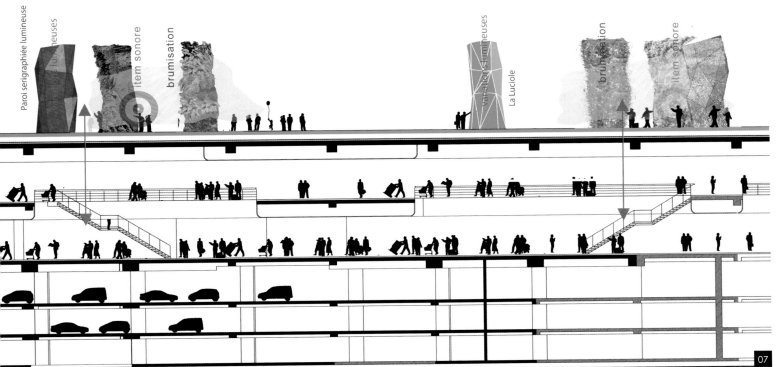

02 公园与公共性花园 Public parks and gardens

塞尔日·盖恩斯伯花园
Serge Gainsbourg Garden
AGENCE TERRITOIRES

地点：法国巴黎
完工日期：2005-2011
面积：2.4 ha
业主：SEMAVIP, 代表巴黎市政府
合作设计师：Gelin & Lafon architectes
照片版权：Nicolas Waltefaugle

Location: Paris, France
Completion date: 2005-2011
Area: 2.4 ha
Client: SEMAVIP, for Paris City Council
Co-project manager: Gelin & Lafon architectes
Photo credits: Nicolas Waltefaugle

01. 整体平面配置图
02. 运动场地
03. 池塘和远处位于环城快速道路出口上的观景平台
04. 日光浴场

01. Overall plan
02. The sports grounds
03. The lake and, in the distance, a view of the Peripheral ringroad
04. Sun-bathing area

在里拉门区段把环城路覆盖起来的做法为巴黎和郊区之间重新建立了联系。在此项目中景观事务所希望在里拉街区和佩-圣-热尔维街区之间建立城际往来，增加一个巴黎大众运输的中转站，安置一个永久性的马戏场，并且在环城路上方设计一个空中花园。

花园是社会与自然共生关系的表达。今天，这种关联存在于生物多样性、生态系统的平衡、资源管理以及减少污染等主题上。在巴黎，意想不到的生物多样性正在环城路沿线的荒废角落里落户生根。这些丰富的花粉媒介促使方案选择采用一些容易在城市环境扎根的旅行植物。花园的轴线沿着环城路而伸展，并以一个望向圣-德尼平原的观景平台作为终点。与这道主要轴线相交的一条羊肠小路侧倚着爬满攀藤植物的石墙而延伸，将不同的活动空间串联了起来。

雨水回收的规划也意味着对地面进行组织并根据需要而赋予适当的材质与形式。位于花园中心的水潭形成一个特殊的生物环境，它同时也收集雨水，补足了一个大型地下水库的蓄水功能，于是花园在灌溉方面得以自给自足。这个覆盖环城路的项目也成为展现历史、地理和生态的绝好时机。城市历史在花园中透过一个能够看到下方环城路的镂空设计而呈现出来：环城路以其既深且广的沟渠形式记录着巴黎旧城墙的历史，也展现出一个边界轮廓清晰可辨的城市。

The covering of the Peripheral ring-road puts the suburbs back into contact with Paris. The Territories agency wanted to reconstruct an urban link with the neighbourhoods of Lilas and Pré-Saint-Gervais, install a city transport link, improve the surroundings for a permanent circus and design a hanging garden above the Peripheral.

The garden expresses the relationship of a society with nature. Today biodiversity, the balancing of ecosystems, the management of resources and the reduction of pollution are the important issues. In Paris, a surprising biodiversity thrives in the abandoned places along the Peripheral. The presence of this rich vector of pollens has led to a choice of imported plants that are capable establishing themselves in the city without difficulty. An axis running parallel to the Peripheral ends in a belvedere overlooking the Saint-Denis plain. A thin walk running beside a wall colonised by lithophyte plants intersects this main axis and feeds off onto the different spaces.

The surfaces have been organised for the harvesting of stormwater. In the centre, the pond, a retention basin for runoff that is developing its own natural environment, works in tandem with a vast underground tank. The garden is autonomous in terms of watering. The project has taken the covering of the Peripheral as a starting point for communicating about history, geography and the living world. The history is revealed through the maintenance of a void where the Peripheral goes through the site. A wide and deep ravine, the space occupied by the Peripheral maintains the memory of the city's fortifications and the idea of a city with limits, marked out and thus identifiable.

05. 池塘周围景观透视图
06. 花园长向剖面图
07. 花园横向剖面图
08. 池塘实景

05. View of the lake
06. General section of the garden
07. Cross-section under the roof of the Peripheral ringroad
08. The lake

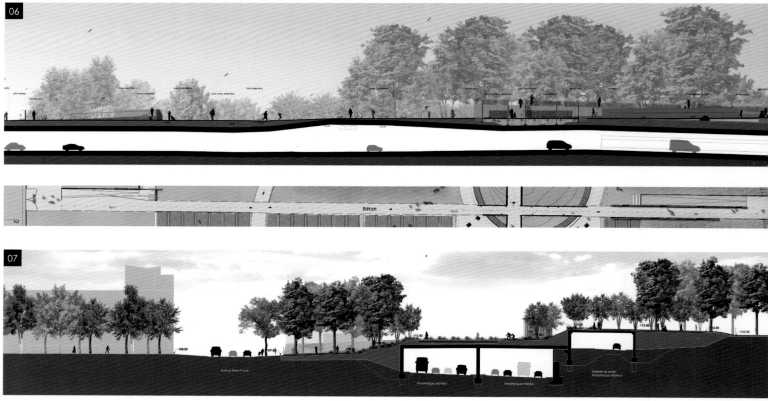

02 公园与公共性花园 Public parks and gardens

09. 中央草坪，一个广阔的空间
10. 池塘上的过桥
11. 细绳般的行人小径，将巴黎和郊区里拉联结起来
12. 望向巴黎的观景台

09. The central lawn, an open space
10. Footbridge over the lake
11. The "thread", a pedestrian pathway between Paris and Les Lilas
12. Viewpoints over Paris

02 公园与公共性花园 Public parks and gardens

风神花园 *Zephyr Garden*

MICHEL CORAJOUD

地点：法国巴黎
完工日期：2007
面积：4.2 ha
业主：巴黎市政府 – DPJEV
合作设计师：Claire Corajoud, ADR Architectes, Stéphane Tonnelat, AEP Normand, ECREP
照片版权：Atelier Corajoud (n°01, 09-13), Jacques Le Bris (n°03-06)

Location: Paris, France
Completion date: 2007
Area: 4.2 ha
Client: Paris City Council – DPJEV
Co-project manager: Claire Corajoud, ADR Architectes, Stéphane Tonnelat, AEP Normand, ECREP
Photo credits: Atelier Corajoud (n°01, 09-13), Jacques Le Bris (n°03-06)

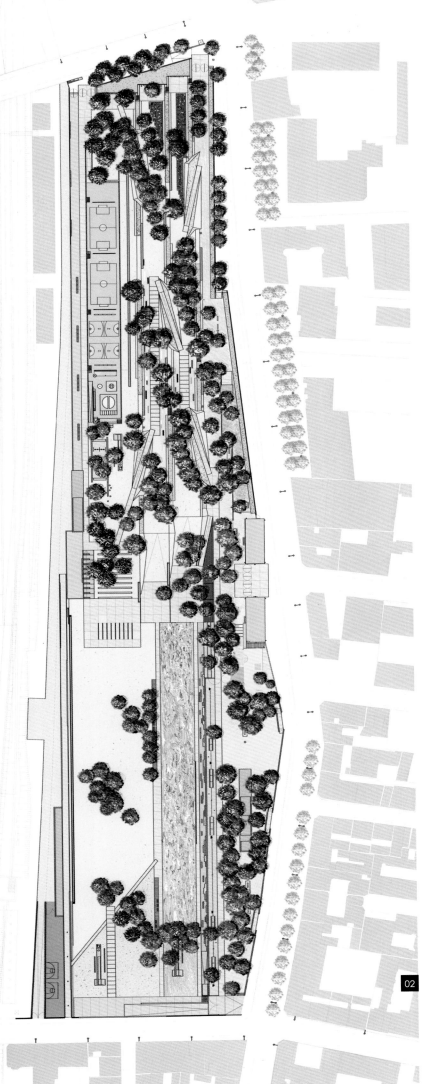

01. 花园全景
02. 整体平面配置图
01. General view of the garden
02. Master plan

摩洛哥宫廷（位于巴黎第18区的表演厅）的新建公园对于一个没有多少绿色空间的街区来说是非常重要的项目，其方案所选定的目标分别为：尊重基地、建构属于日常生活的空间、创建多样化的生态环境、实施公园的生态化管理、规划光线和阴影、遵循建筑高质量环境的规范标准。

在纵向组织的方式下，这个取名为"风神花园"的摩洛哥宫廷公园由两部分组成。在南边，花园从东向西分成六个部分：观景广场、木板岸台、运河、砾石花园、大草坪、石栏围墙。另外还有南侧角落坡道、青少年游戏场、栈桥、托儿所花园。北侧，花园的特色展现于鲜明的起伏地势，形成三层连续的平台：一个长生植物花园、一个禾本植物花园和一个运动平台。这些平台被一条林荫道、一个共享花园和一条连廊所环绕。

Morocco Court Park is a vital project for a neighbourhood that lacks green spaces. Its aims are: respect for the site, a project for neighbourhood life, the creation of varied ecological environments, the ecological management of the park, management of light and shadows both day and night, and a High Environmental Quality (HQE) approach for the buildings.

Arranged lengthways, Morocco Court Park, which is known as the "Zephyr Garden", breaks up into two parts. In the south, the garden is organised from east to west into six strata: the esplanade, the wooden deck, the canal, the gravel garden, the large meadow, the screen wall. To this are added the ramp in the south corner, the teenagers' playground, the footbridge, the crèche garden. In the north, the garden's relief is pronounced. It rises in three successive terraces with a perennials garden, a grasses garden and a sports terrace. These terraces are "bordered" by a shady walk, a garden for local non-profit associations, and a walkway.

03. 大草坪，其一旁沿靠着栈桥
04. 石栏围墙确保了公园与铁道之间的视觉穿透性
05. 木板岸台居高临下眺望运河和砾石花园
06. 大台阶可以连通运动平台的观景台
07. 08. 大广场和栈桥之间的横向剖面图

03. The large meadow, extended by a footbridge
04. High wall ensuring visibility between the park and the railway
05. The wooden decked quay overlooks a planted canal and a gravel garden
06. Grand staircase leading to the belvedere of the sports terrace
07-08. Cross-sections between the esplanade and the footbridge

这南北两个部分通过一个大台阶和几个花园相互连接，这些花园面向一个大广场和一片人工雾景。这个公园的步行功能和多样化活动使得灯光环境的构思成为必要的一环，即使在夜晚也要充满活力且变化多端。

环保概念被纳入到项目的每个建造阶段中，当然也贯穿于植被的整个生长过程中。在此框架下，民选代表、联合会、周边居民、技术服务部门等全部联合起来参与方案的构思，巴黎市政府同时希望能有社会学家的参与，以便更好地把项目相关者的期望都整合其中，尤其是周边街区居民的需求，使他们能够预先对新公园产生归属感而善加使用。

These two parts are linked via a wide staircase and tiers, leading to the large squares and an artificial mist feature. The mainly pedestrian function of the park and the diversity of its activities call for dynamic and varied luminous environments, even at night.

Each phase of the project has been approached with a commitment to the environment, and of course that includes the life of the garden. Local politicians, associations, inhabitants and technical services have worked together to this end. Paris City Hall asked for a sociologist to be involved in order to make sure that the various parties' aims and wishes were taken account of and, above all, those of the local inhabitants, thus allowing them to get a feeling of ownership of this new park in advance.

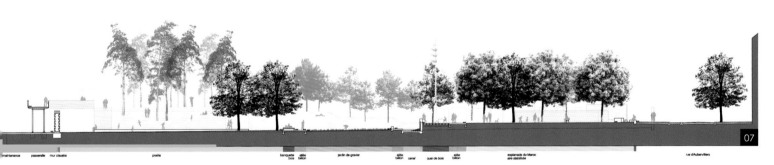

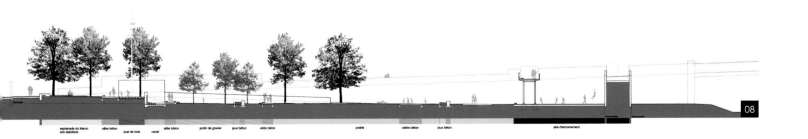

02 公园与公共性花园 Public parks and gardens

09. 木板岸台上设置了长条桌椅
10. 运动平台
11. 公园里的三个连续平台绝大部分覆盖着草皮和树木
12. 种植着水生植物的水渠提供了一个出色的生物群落
13. 灌木丛和树木将这个大草坪分成三个空间

09. Long tables are installed on the wooden quay
10. The sports terrace
11. The terraces are mainly laid to lawn and covered with trees
12. The canal planted with aquatic plants offers a remarkable biotope
13. This wide meadow is divided into three, with the aid of tree copses

02 公园与公共性花园 Public parks and gardens

塞甘岛预示花园

Seguin Island Prefiguration Garden

MICHEL DESVIGNE PAYSAGISTE

地点：法国布洛涅-比扬古
完工日期：2010
面积：2.3 ha
业主：塞纳河谷混合经济开发公司
照片版权：Michel Desvigne Paysagiste

Location: Boulogne-Billancourt, France
Completion date: 2010
Area: 2.3 ha
Client: SAEM Val de Seine Aménagement
Photo credits: Michel Desvigne Paysagiste

01. 方案位置与整体配置图
02. 空间整治随着时间而演变
03. 岛上第一个开放的公共空间
04. 未来整治空间的原型和预示

01. Situation plan
02. An evolution that grows with the development of the site
03. The first public site to open on the Island
04. Prototypes and prefiguration of the future development

塞甘岛的花园是岛上第一个开放的公共空间，这个占地2公顷的花园将是改造后的岛屿中央花园的前身。这个具有预示性的花园将作为塞甘岛改造项目的基础，伴随着工程进展而逐步演变。

花园的构图呼应着雷诺工厂的历史印记。其硬质铺地呈现出昔日设置着一个个压力洗车地沟的广大混凝土基座的痕迹。以混凝土和稳固沙子塑造而成的长方形平台，组织着一系列高低不同、铺着草皮或长满植被的空间。成排的小柳树为植物群建立起架构，其间则种植着先锋植物。稍后这些临时性植物将被持久性的植被所取代，以适应周围的新城市环境。

这个花园是观察小岛上的工地以及四周变化中的土地的绝佳场所，并且以充满趣味的方式把它周围正在建造的设施展现出来：广大的儿童游戏沙坑、实验性的分享花园（家庭式花园）、野餐草地、餐厅⋯⋯ 然而这个花园的主要功能还是在于为公众提供一个了解和进入塞纳河谷核心区的窗口。

由建筑师伊内萨·安驰所设计的城市小品以线状形式分布在基地上。由艺术家彦纳·科尔萨列所设计的照明设备以镀锌钢柱廊的形式呈现，形成高层的空间元素。

The garden is the first public space to open on the Ile Seguin, and its two hectares are designed to give a foretaste of the garden that will take centre stage when the island has been developed. This prefiguration is a foundation that will evolve as the development unfolds.

The geometrical composition of the garden plays with the memory of the Renault factory. The hard surfaces follow the lines of the solid concrete base once punctuated by the pits of the body presses. Simple rectangles of concrete and stabilised sand organise the succession of spaces on different levels where lawns and plantations grow. Rows of small willows give it a structure. Pioneer plants have been established between these lines. This temporary vegetation will later be replaced by perennial plants adapted to the new urban environment.

This garden has the privilege of being the observatory for the island building site and this changing land area. In a playful way it sets a scene for the building site that surrounds it: huge sandpits for children, experimental gardens for local associations, meadows for picnics, a restaurant... But above all, it is an open place offered to the public at the heart of the valley of the Seine.

The furniture designed by the architect Inessa Hansch is arranged in long lines according to the overall design of the site. The lighting, designed by the artist Yann Kersalé, uses galvanised steel arches that form a high strata

05. 城市小品随着基地尺度而呈长线形排列
06、07. 由建筑师伊内萨·安驰所设计的城市小品

05. Furniture arranged in long lines according to the configuration of the site
06-07. Urban furniture by Inessa Hansch

02 公园与公共性花园 Public parks and gardens

08、09. 雷诺汽车工厂旧基座的重新诠释
10、11. 观察站让人得以对这个变化中的大地景观进行了解
08-09. Reinterpretation of the site of the old Renault factory
10-11. An observatory for a land area undergoing change

02 公园与公共性花园 Public parks and gardens

普朗诗岛屿溢洪道公园
Planches Island Park

HYL

地点：法国勒芒
完成日期：2009
面积：3 ha
业主：勒芒市政府、勒芒大都会区理事会
合作设计师：Arnaud Yver architecte
照片版权：HYL

Location: Le Mans, France
Completion date: 2009
Area: 3 ha
Client: Le Mans City Council, Le Mans Métropole
Co-project manager: Arnaud Yver architecte
Photo credits: HYL

01. 整体平面配置图
02. 从高处走下溢洪道
03. 绿地剧场和溢洪道
04. 淹水时候的溢洪道

01. Master plan
02. Pathway to the weir
03. Amphitheatre and weir
04. Flooding of the weir

位于勒芒市正中心位置的一个旧工业岛屿因为获得一个全新功能而得以重新转变成一片自然之地。这个新功能即是：成为一个"水利调节设施"，在涨水期作为萨尔特河以及其运河之间的溢水系统。

方案必须尽可能地把被溢洪道分割成的零落空间、三座桥、不协调的河岸以及一个小型房地产项目等元素结合起来。设计师决定通过石砌护坡、加固路堤、砖砌防护墙等设施，来将岛屿的主要空间规划为观景平台。当溢洪道的技术功能被暂置一旁的时候，这个制高点便犹如与城墙对话的堡垒，它也变成在一个绿地剧场和一条长长散步坡道上所凝望观赏的舞台景观；因为剧场和散步道同时化身为日光浴场，并接纳来草地上野餐的人们。靠近运河的一边，河岸的缓坡以天然状态呈现，并柔和地过渡到水边，无需任何围栏的护卫。

普朗诗岛屿公园犹如一座由砖块、石头和黏土砌成的河畔小城堡，它重新为这个河流汇流地带来地形上的统合，并且在城市中心缔造出新的景观视野。方案实施之后，岛屿上的桥梁拉近了它与人们的距离，它也成为那些和谐融入萨尔特河景观中的几个街区的"中心点"。

In the centre of Le Mans, an industrial island has become one with nature again... thanks to its new function as a "hydraulic machine": serving as an overflow weir between the River Sarthe and its canal during high waters.

As far as possible HYL had to unify the space that was fragmented by the weir, three bridges, disparate banks and a small housing operation. The plan was to turn the main part of the island into a lookout point through the use of riprap, strengthened embankments and brick supports. This high point becomes a kind of bastion creating a dialogue with the walls of the housing estate, while the weir forgets its technical vocation and becomes the setting for a grassy amphitheatre and a long ramp-walk where people come to sunbathe and picnic. On the canal side, the slope of the bank that has been left "natural" is softened to allow people to look down to the water without any guardrail.

A small river citadelle made of brick, stone and earth, Planches Island park gives the confluence back its geographical unity and provides new, attractive views for the housing estate. With its bridges making it far more accessible, the island has become the umbilicus of several neighbourhoods cohabiting harmoniously in the federating landscape of the Sarthe.

05. 城市中的公园
06. 萨尔特河畔的堤岸平台
07. 绿化剧场的阶梯
08. 秋季的溢洪道
09. 绿色休憩平台

05. A park in the city
06. Terrace quay on the Sarthe side
07. Amphitheatre steps
08. View towards the weir in autumn
09. The green terraces

02 公园与公共性花园 Public parks and gardens

马尔贝街区公园散步道
Marbé Neighbourhood Park-Promenade

IN SITU

地点：法国马孔
完工日期：2013
面积：35 000 m²
业主：马孔镇政府
照片版权：In Situ (n°04, 06-14), Patrick Georget (n°02, 03)

Location: Mâcon, France
Completion date: 2013
Area: 35 000 m²
Client: Mâcon Town Council
Photo credits: In Situ (n°04, 06-14), Patrick Georget (n°02, 03)

01. 马尔贝街区平面配置图
02. 蛇状的小径
03. 运动场地和新开设的横向街道

01. Master plan of the Marbé neighbourhood
02. The Serpentine walk
03. The sports field and new cross-paths

在马孔的马尔贝街区、采石场街区和德塞尔街区的中心，继几栋板楼和塔楼被拆除之后，一条新规划的公园散步道沿着与索恩河平行的550米长空间而延伸、发展。这条蜿蜒曲折、灵活多变的宽阔路线成为不同街区间既具有社会性又具有使用功能的纽带。公园内铺展着一束束波浪般摇摆的小路，它们既作为步行道，同时也是自行车道，周围植被茂盛而构成"肌束"般的条纹，成为街区中心的一片绿色田野。

马尔贝街区的公园散步道盘绕着几乎已被遗忘的阿比末溪流，其河床很深且难以看到，直到方案实施之前都是完全无法靠近的。以阿比末溪流作为方案构思原则的原因来自于让人们接近水流的愿望，此构想的实现必须考虑到与涨潮相关的一些限制条件。一些阶梯的设置便于人们接近溪流，两个栽培了大量胶须藻的池塘也依靠阿比末溪的水来进行补给。

Following the demolition of several housing blocks and towers in Mâcon, the park-promenade runs along almost 550 metres parallel to the Saône at the heart of the neighbourhoods of Marbé, Les Perrières and La Desserte. It is a wide, flowing, winding route that provides a social and functional link between the different neighbourhoods. The park unfolds in an array of undulating paths shared by pedestrians and cyclists and forms a plant "muscle", a furrow at the heart of the neighbourhoods.

The Marbé park-promenade follows the Abîme brook, a forgotten but unusual element. The riverbed is deep and hard to see, and was until present totally inaccessible. The Abîme project's principles are based on the wish to get closer to the water while integrating the constraints linked to flooding. Terraces allow one to get down to the brook and the two river pools are abundantly planted and fed by the water from the river.

04-06. 两旁设置着雨水花园的散步小径
04-06. The promenade walks bordered by storm gardens

往北延伸的公园被拉展成沟脊状，蜷卷在条状的散步道之间。这两条小径向两侧分开，中间设置了儿童游戏场、点缀着鲜花的广阔草场以及一层层皱褶般的草丛，每每经过街道便又汇合在一起，以人行小平台的方式穿越道路。游戏设施背靠着成片的欧洲赤松，水生花园则伴随着家庭式花园。这些家庭式花园成为新住宅街区边缘的活动场所，也是位于私人住宅和公共公园之间具有实用功能的绿地。在整个公园散步道上，横向贯穿的小路将公园与附近的街区和戴高乐大道连接了起来，不仅确保了马尔贝公园散步道和索恩河之间人行空间的连续性，也为两者建立了真正的联系。

Going north, the park spreads out in the form of furrows and ridges tucked in between the ribbons of the walk. The two paths diverge, leaving space for playgrounds, wide flower meadows and rolling folds of grass, and come together again to the right of the roads, crossing them on a same-level pedestrian plateau. The playground structures back onto to hillsides planted with sylvester pines, while the wetland gardens are next to the family vegetable gardens. These allotments form an active fringe around the newly restructured residential blocks. It is a place of with a double use – by the private residences and as a public park. All along the park-promenade cross-paths open the park out to the neighbourhood and Avenue de Gaulle, forming real pedestrian links and continuity between the park-promenade of Marbé and the Saône.

02 公园与公共性花园 Public parks and gardens

07. 位于公园边缘的家庭式花园
08. 这些葡萄藤架同时也成为花园的庇护所
09. 阳光普照的小径
10. 雨水收集以便灌溉花园地块

07. Park side: edge of the allotments
08. Vines and garden shade
09. The sunny walk
10. Rainwater harvesting for watering the allotments

02 公园与公共性花园 Public parks and gardens

11. 水泥块铺成的穿越小径
12. 运动场地和雨水花园
13. 阿比姆公园
14. 边缘设置石笼的公园小径

11. A concrete cross-path
12. Sports field and storm gardens
13. Abîme park
14. The park walk bordered by gabions

02 公园与公共性花园 Public parks and gardens

东印度公园
Park of the West Indies
IN SITU

地点：法国洛里昂
完工日期：2014 – 第一阶段
面积：5.6 ha
业主：洛里昂镇政府
合作设计师：Atelier CMJN architectes
图片版权：Les Éclaireurs (n°01), In Situ (n°02, 04), Atelier CMJN (n°03, 05, 06)

Location: Lorient, France
Completion date: 2014 – 1st phase
Area: 5.6 ha
Client: Lorient Town Council
Co-project manager: Atelier CMJN architectes
Image credits: Les Éclaireurs (n°01), In Situ (n°02, 04), Atelier CMJN (n°03, 05, 06)

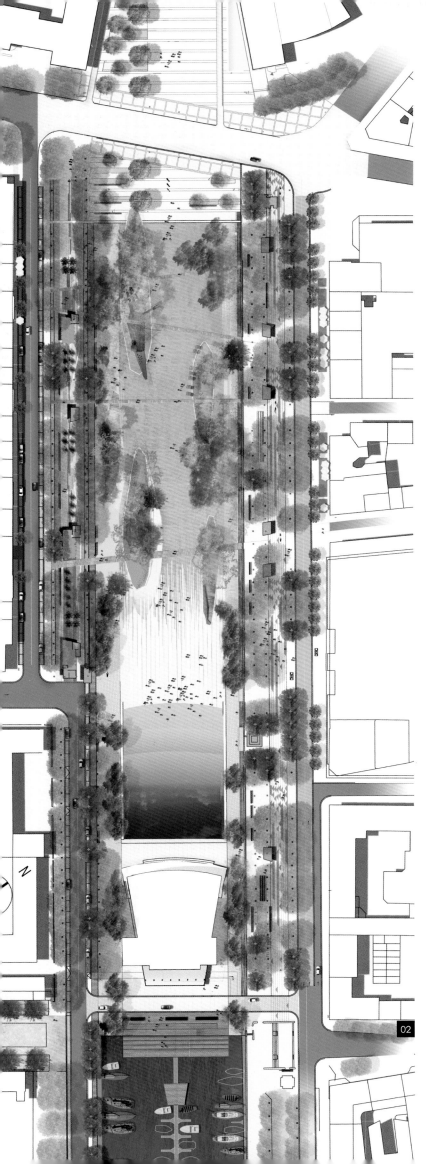

01-02. 从港口延伸出来的混合型公共空间
01-02. A hybrid public space as a continuation of the port

儒勒·菲里广场是位于市政厅、港口、印度堤岸和罗昂堤岸之间的一个大型公共空间。如此一个位于市中心、占地5.6公顷的广阔空间，在发展策略上具有相当地理优势。然而，这个场所已经逐渐变得老旧不堪：两侧河堤上的停车场、植被、被阻隔的视线和受阻碍的步行道，这些都为两岸形成了诸多隔阂。这片土地需要重新被注入活力以便成为一个活跃的公共空间，一个以历史和洛里昂地区特征作为基础的崭新公园"东印度公园"。

尹西图事务所的方案使广场脱离原先的闭塞状态，在港口的延展线上建设一个面向大海的中央公园。它既是广场也是公园，强化了公共空间的延续性和统一性，并展现出洛里昂的海滨特色。印度堤岸成为一条真实的热闹大街，伴随着一条公共交通专用车道和各种软性交通（步行、自行车、滑轮……）通道，沿途设置有售货亭和休息平台。公园里开设了一片广阔的草坪，并在其中排除了所有硬质铺地和不渗水的地面，因此显得既简洁又丰沃。零星点缀其中的"花园岛"上容纳着各式各样的迷你植物园、游戏场、休憩空间……

Jules Ferry Square forms a large public space between the Town Hall and the port, flanked by the Quay of the Indies and Rohan Quay. Its strategic situation is exemplary: a huge space of 5.6 hectares is available in the heart of the town centre. But the place has aged badly: the parking on the two quays, the planting, the visual and pedestrian ruptures all form barriers on the two sides. This land area just needs to be reactivated to become a dynamic public space, the new "Park of the West Indies" in reference to Lorient's history and identity.

In Situ's project opens up the square and expresses Lorient's maritime history by strengthening the continuity and the unity of a federating public space in continuation of the port: a purposeful "central park" that is both square and park, overlooking the sea. The Quay of the Indies becomes a real "rambla", bringing together a lane for the exclusive use of public transport, and soft transport lanes punctuated by kiosks and terraces. The park is simplified and fertilised by creating a vast meadow and getting rid of all the hard surface and non-porous zones. The "garden islands" that punctuate the meadow host different botanical microcosms, playgrounds, areas for relaxation, etc.

03. 位于会议中心脚下的潮水广场
04. 横向剖面图：公园、潮水广场和港口
05. 东印度堤岸成为一条生动活泼的大街
06. 潮水广场，一个可吸收洪泛的空间

03. Place des Marées, at the foot of the Palais des Congrès
04. Cross-section of the park, Place des Marées and the port
05. Quay of the West Indies, already in use as a "rambla"
06. Place des Marées, a floodable public space

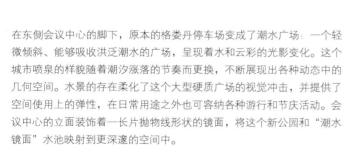

在东侧会议中心的脚下，原本的格娄丹停车场变成了潮水广场：一个轻微倾斜、能够吸收洪泛潮水的广场，呈现着水和云彩的光影变化。这个城市喷泉的样貌随着潮汐涨落的节奏而更换，不断展现出各种动态中的几何空间。水景的存在柔化了这个大型硬质广场的视觉冲击，并提供了空间使用上的弹性，在日常用途之外也可容纳各种游行和节庆活动。会议中心的立面装饰着一长片抛物线形状的镜面，将这个新公园和"潮水镜面"水池映射到更深邃的空间中。

In the east, at the foot of the Palais des Congrès, the current Glotin car park will become Place des Marées: a floodable square whose slight incline allows one to work with the effects of water and sky. This urban fountain plays with the rhythm of the sea and offers a space with a variable geometry that is constantly changing. The presence of the water softens the impact and the way this wide hard-surfaced esplanade is used, and provides the flexibility needed for hosting events and the interceltic festival. The facade of the Palais des Congrès, clad in a long parabolic mirror, has a plunging view over the new park and the "mirror of the tides" ornamental lake.

02 公园与公共性花园 Public parks and gardens

彼 岸 *The Other Bank*
CATHERINE MOSBACH

地点：加拿大魁北克
完工日期：2008
面积：150 m²
业主：Espace 400°
照片版权：Catherine Mosbach (n°02-09, 12-14), Patrick Rimoux (n°01)

Location: Québec, Canada
Completion date: 2008
Area: 150 m²
Client: Espace 400°
Photo credits: Catherine Mosbach (n°02-09, 12-14), Patrick Rimoux (n°01)

01. 冬天，2010 年
02. 秋天，2011 年
03. 春天，2008 年
04. 灌木丛，2008 年

01. Winter 2010
02. Autumn 2011
03. Spring 2008
04. 2008 bushes

"彼岸"项目的抽象环境扮演着方法论的角色，因为它象征着从一个耕作出发抵达它处的可能性。这里"耕作"一词意指就地准备土壤以便植物扎根。它同时也用于比喻，让原本不存在的展现出来：光线透过棱镜显现的色差、树林的弹性、树汁的力道、树叶的闪光、花序的柔软、土壤的多样形态、光线的能量。

环境形成限制，但同时也展现出由一种风格所引发的各种感知，并且让一种形态和其背景变得具体可触及。它使造访者置身于他所发现的意象和他已经知晓的影像之间。材质的密度展现出尺度，瓦解对已知物与肯定性的揣测。柳条细枝的可塑性在液体与固体之间、在静止与活力之间、在光线与色彩之间塑造不同的差距。水的旋动与枝干的弹性。

插条占据着基础与支架，此乃长出枝条的原始状态。它与树根一起展现出之前与之后、起始与结束，展现出所有诠释的无限活力。

The abstract framework of the Other Bank plays the role of instrument, of a time signature that suggests the possibilities of an elsewhere from the strands of a culture. The word culture is used in its true sense, as in preparing an existing soil for the planting of root stock. It is also used in the figurative sense, as in making appear what is not already there: the chromaticism of the light prism, the flexibility of wood, the power of the sap, the luminescence of leaves, the velvetiness of catkins, the polymorphism of a soil, the energy of light.

The framework limits, but at the same time it opens up the understanding of a composition and makes tangible a pattern and its background. It places the visitor in a layer between the visions that he discovers and the images that he already knows. The density of the materials distils the scale, dissolves the measuring of acquired knowledge, of certitudes. The plasticity of the wicker branches works the margins between liquid and solid, between inert and living, between light and colour. The currents of water and the flexibility of stems.

Hardwood cuttings occupy the base, the underlying structure, the initial state from which the wicker rises. With the root stock it introduces the before and after, the opening and its ending, vitality in all interpretations.

05. 秋天，2011 年
06. 林冠，2008 年
07. 捕捉，2008 年
08. 新叶，2008 年
09. 装饰，2008 年
10. 节外生枝，竞赛图案
11. 移植栽种，竞赛图案

05. 2011 autumn landscape
06. 2008 canopy
07. 2008 plantings
08. 2008 foliage
09. 2008 in its finery
10. Outgrowth, competition drawing
11. Transplanting, competition drawing

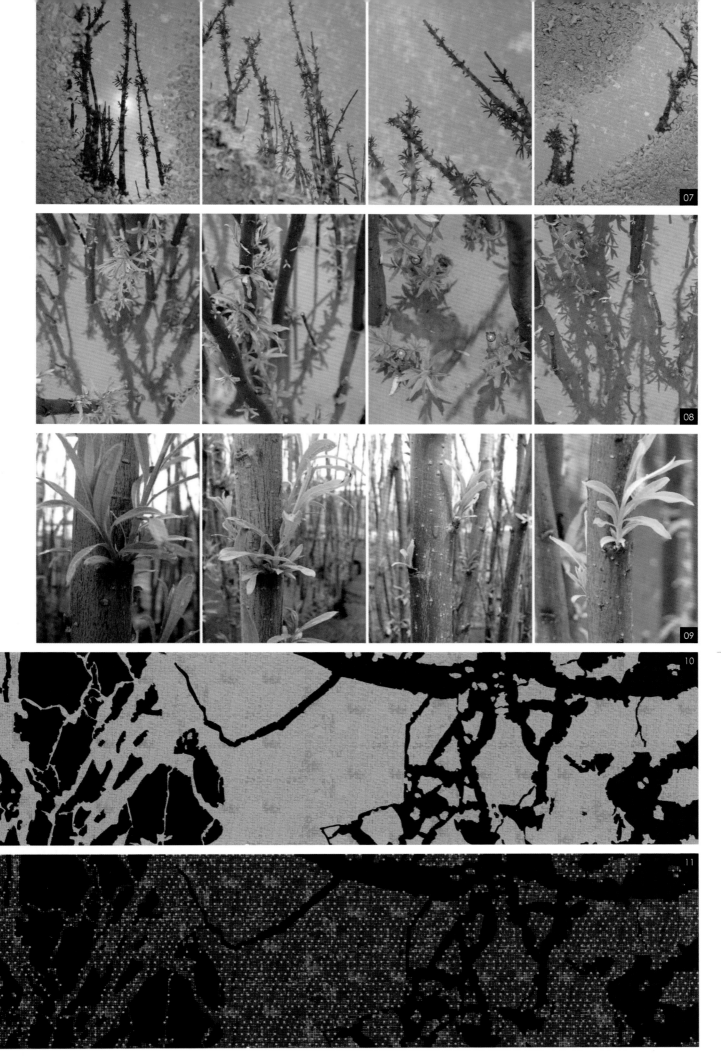

02 公园与公共性花园 Public parks and gardens

12. 当年的成长
13. 显露
14. 植物如丛林般向下扎根

12. The year's new growth
13. Emerging
14. The plants take root

02 公园与公共性花园 Public parks and gardens

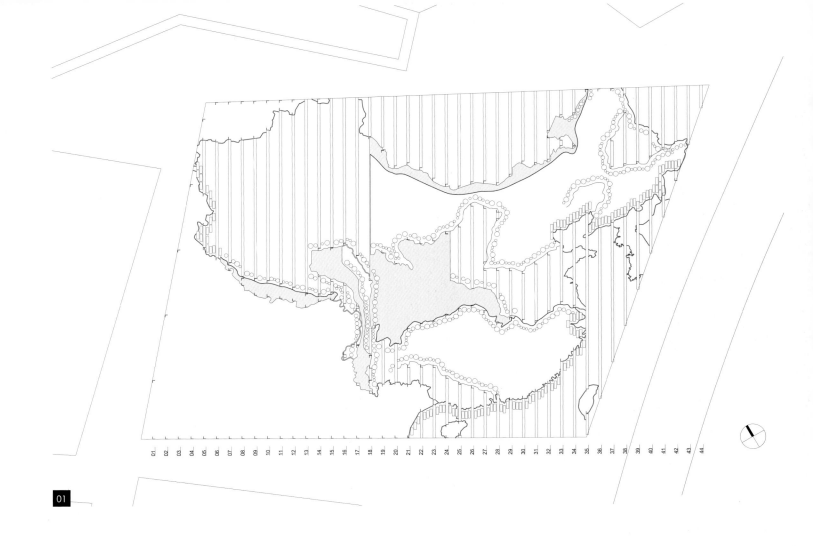

01

SPOT5 花园 / 山水意象
Spot5 / Shanshui "Mountain Water"
CATHERINE MOSBACH

地点：中国西安
完工日期：2010-2011
面积：1 000 m²
业主：IHE (国际园艺展览组织单位)
照片版权：Catherine Mosbach

Location: Xi'an, China
Completion date: 2010-2011
Area: 1 000 m²
Client: IHE (International Horticulture Exhibit)
Photo credits: Catherine Mosbach

01. 平面图：水景-过道-门廊
02. 两者之间
03. 穿越

01. Water-pathways-steel profiles
02. In-between
03. Pathways

位于西安的Spot5能够举办各型大小活动，邀请造访者超越场地边界的限制而沉浸在一个微型宇宙之中。方案形貌主要以点与线进行最简约的表达：前者在空间里推进，以呈现流动的几何形式，并成为大地广阔尺度与细胞微观尺度之间的转口；后者则呈现距离，使事物维持一体的最小间距。两者一起表达出流动与迁徙的结合，勾勒出生物的形成，犹如集体空间的形成。

一系列的线性结构重现出中国地图，其内部构图参考了中国领土与历史上多处存在的长线形建构物在大自然环境中分布和自我保护的方式。

遵循着中国皇家园林的规则，并透过南北向线性框架的转置，此花园平面呈现出中国领土轮廓与其十大河流的象形图案。花园展现在这一连高低不等的串框架下，造访者在此随着季节变化而发现各种独特的景象：叠立在蓄水池之上的框架、淹没在盛开植物丛里的框架……这些框架的组合使得花园能够以多样面貌呈现，从水中倒影的一个细节到随风摇曳、多彩多姿的繁茂植物。

Spot5 in Xi'an is a location for events both large and small, and invites visitors to submerge themselves in a microcosm that goes beyond the limits that shape it. Lineaments have been favoured, crafted in a simple way like dot-to-dots. One evolves in the space to represent the geometry of movements and journeys from the large scale of the land area down to the microscopic scale of cells. The other marks the distances and small inter-distances that cause things to hold together. Both translate the combinations of movements and migrations that draw the form of the living and of shared spaces.

The transcript grid here is a cluster of linear structures, echoing the infinite linear works present everywhere on the land area and in the history of China in the way that they ramble over and/or protect great natural sites.

Following the precepts of the great imperial parks, Catherine Mosbach used a pictogram showing the contours of the country, engraved by its ten main rivers, transposed by an array of lines running from north to south. The garden takes form from this collection of silhouettes, from which the visitor discovers unusual scenes according to the season: naked silhouettes overlooking a water retention lake, or silhouettes semi-submerged in a mass of vegetation. The combinations of these compositions of silhouettes in their settings offer innumerable possibilities for garden scenes ranging from the detail of a reflection in a puddle to the fluctuating coloured masses waving in the wind.

04、05. 等角透视空间研究
06、08. 穿越

04-05. Axonometric studies
06-08. Pathways

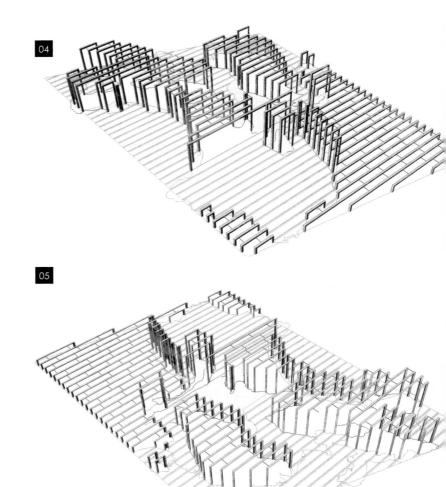

147

02 公园与公共性花园 Public parks and gardens

游戏场公园
Playground Park

OLM / PHILIPPE COIGNET

地点：法国埃尔蒙
完工日期：2013
面积：1 ha
业主：埃尔蒙镇政府
合作设计师：NStudio Architectes (设计总负责)
图片版权：OLM

Location: Ermont, France
Completion date: 2013
Area: 1 ha
Client: Ermont Town Council
Co-project manager: NStudio Architectes (project representative)
Image credits: OLM

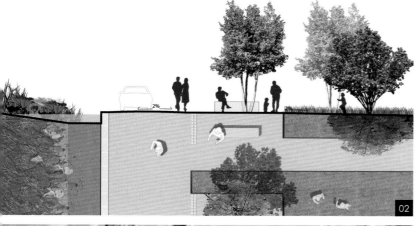

01. 弗朗索瓦·吕德社会文化中心透视图
02. 渗透池剖面平面图
03. 整体平面配置图
04. 游戏场的三项组织性原则

01. View towards the François Rude sociocultural centre
02. Cross-section of the infiltration basin
03. Master plan
04. Three structuring principles of the playing field

游戏场公园项目占地1公顷，其整治目标是将一个现存的游戏场改造成为一个配有新的社会文化中心的运动公园。位于罗兰·加洛斯中学下方一个质量低下的环境中，现有的方案基地容纳了两个尺度不当的水泥广场。方案利用基地多变的地形：沟渠、平地、斜坡和低地等，塑造出文化、运动和娱乐等不同场所，建立了一个多功能的城市公园。

一条宽大的中心通道将建筑与公园连接起来，并通过由经过稳固处理的沙土小路所组成的网络联系了北部的三个游戏场。为了适应某些体育活动的需求，方案重新利用并拓宽了原有的水泥场地，这些地面的高度被重新调整，以便将雨水引到路边的壕沟中，收集到的雨水将用于浇灌公园边界的树木以及隔离每个游戏场的灌木树篱。

在中心通道的南边，一片大草坡坐落在蓄水池上方，水池收集了社会文化中心的屋顶雨水和一部分基地的地面雨水。没有做防水处理的水池不仅缓缓地向地下含水层渗入水分，也为水池中的亲水性植物提供水源，这些植物见证着水池的水量升降。

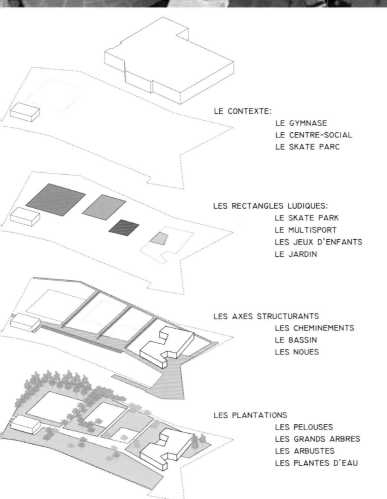

LE CONTEXTE:
    LE GYMNASE
    LE CENTRE-SOCIAL
    LE SKATE PARC

LES RECTANGLES LUDIQUES:
    LE SKATE PARK
    LE MULTISPORT
    LES JEUX D'ENFANTS
    LE JARDIN

LES AXES STRUCTURANTS
    LES CHEMINEMENTS
    LE BASSIN
    LES NOUES

LES PLANTATIONS
    LES PELOUSES
    LES GRANDS ARBRES
    LES ARBUSTES
    LES PLANTES D'EAU

The one-hectare project consists of transforming an existing playground into a sports park with a new socio-cultural centre. The existing site offers two small concrete surfaces in an unremarkable space below the Roland-Garros High School. The project aims to offer an urban park with distinct functions organised via a topographical arrangement of ditches, raised surfaces, slopes and dips that structure the different cultural, sports and recreational facilities on offer.

A wide central walk links the building to the park and feeds the three playgrounds in the north via a network of paths in stabilised soil. The concrete surfaces are re-used and enlarged to be used for several sports activities. They are levelled to send stormwater into large ditches running alongside the paths, to be absorbed by the stem trees on the edges of the site and by the shrub hedges that separate the playgrounds.

South of the central walk, a large grassy slope overlooks an infiltration basin for harvesting stormwater from the roof of the socio-cultural centre and runoff from part of the site. The permeable basin slowly replenishes the groundwater and supports hydrophilic plants which show the volume of water present in the basin.

Martin Luther King Park
马丁·路德·金公园
ATELIER JACQUELINE OSTY & ASSOCIÉS

地点：法国巴黎
完工日期：2007 – 第一阶段
面积：10 ha (第一阶段4.4 ha)
业主：巴黎DEVE
合作设计师：François Grether architecte (设计总负责)
照片版权：Paris Batignolles Aménagement (n°01), Arnauld Duboys Fresney (n°03, 04, 06-13), DEVE (n°02), AJOA (n°5)

Location: Paris, France
Completion date: 2007 – 1st phase
Area: 10 ha (4.4 ha 1st phase)
Client: DEVE Paris
Co-project manager: François Grether architecte (project representive)
Photo credits: Paris Batignolles Aménagement (n°01), Arnauld Duboys Fresney (n°03, 04, 06-13), DEVE (n°02), AJOA (n°5)

01. 鸟瞰街区全景透视图，公园融入街区之中
02. 秋天公园全景
03. 公园景观
04. 池塘提供了生物群落环境

01. Aerial view showing how the park fits into its neighbourhood
02. View of the whole park in autumn
03. View over the park
04. The biotope basin

克利希-巴蒂诺尔新街区位于一块40公顷的废弃铁路遗址上，其中10公顷土地被预留作为马丁·路德·金公园：新的城市形态正是围绕着这个中央空间形成的。作为街区组织的结构性元素，公园与各种交通流线之间的通达性成为设计的重点。同时，为了更好的连接位于贝雷尔的云杉社区和位于克利希城门的巴蒂诺尔街区，周围的城市轴线以两侧植树的林荫道形式，延伸到公园内部。

在景观方面，设计师融合了不同的设计手法，以增强大自然的效果。为了使不同季节从北到南依次展现各自的魅力，季节更迭成为方案的植物种类选择的依据：丰富娇艳的花朵在春天绽放；光与影、草坪和禾本植物在夏天嬉戏；闪亮的枝叶在秋天舞动；苍劲树皮、桦树和松树则留给冬天。水则透过几种不同的形式而展现：技术性的、环境的和趣味性的。虽然水体是借助人工的方式流动，但其景观质量与生态属性仍然完美地融合在一起。观赏池塘形成一个个生物小环境，生长着典型的湿地植物。经过绿化的美丽沟渠收集着雨水，以便用于灌溉公园。观赏和游戏喷泉的水都可以一再循环使用。

就使用功能来说，公园通过一些大型开放空间、若干私密场所以及作为儿童游戏场或者运动设施、表达着新城市美学的带状空间，为人们提供了相当多样化的活动空间。公园的高度使用频率证明了它对当今人们的需求与期望作出了极佳的回应。

The new neighbourhood of Clichy-Batignolles covers 40 hectares of abandoned railway land, of which 10 hectares were reserved for the Martin Luther King Park. The new urban forms are organised around this central space. Because the park has a structuring role for the neighbourhood, its accessibility for those wishing to use it to get from one place to another is a marked characteristic. The urban lines around it extend into the interior of the park in the form of planted walks that link the neighbourhood of Épinettes to that of Péreire, and Batignolles to Porte de Clichy.

In terms of landscape design, different devices are employed to amplify the effects of nature. The cycle of the seasons guided the choice of plants so that from the south to the north the seasons can be seen at their best as they follow on from each other: abundant flowerings in spring; a play of shadows and light, lawns and grasses for summer; resplendent foliage in autumn; bark, and copses of birches and pines for winter. Water is present in several forms: technical, environmental and playful. Even when its flow is artificially controlled, its landscape qualities are combined with its ecological properties. The ornamental lakes are biotopes filled with a vegetation symbolic of wetland environments; the pretty planted ditches harvest stormwater for irrigating the park; the water in the fountains is recycled.

When it comes to uses, with its wide open spaces and its more intimate corners, its active strips dedicated to children's playgrounds and sports facilities that define a new urban aesthetic, the park offers a wide variety of activities. And its popularity proves that it is in tune with the expectations and desires of today's public.. This project, with the City of Paris, won the Special Prize "When the garden builds the city" at the Victoires du Paysage 2012.

05. 春天的植物沟渠
06. 秋天的植物沟渠与风车
07. 水池畔
08. 夏季公园景观

05. The wetland ditch in spring
06. Windmill and the wetland ditch in autumn
07. Beside the basin
08. The garden in summer

02 公园与公共性花园 Public parks and gardens

09. 公园里的带状活动空间
10. 水柱广场
11. 儿童游戏场的带状活动空间
12. 公园小径
13. 线形花园

09. The park's activity stripes
10. The fountain square
11. The activity stripes of the children's playground
12. A park walk
13. The linear garden

02 公园与公共性花园 Public parks and gardens

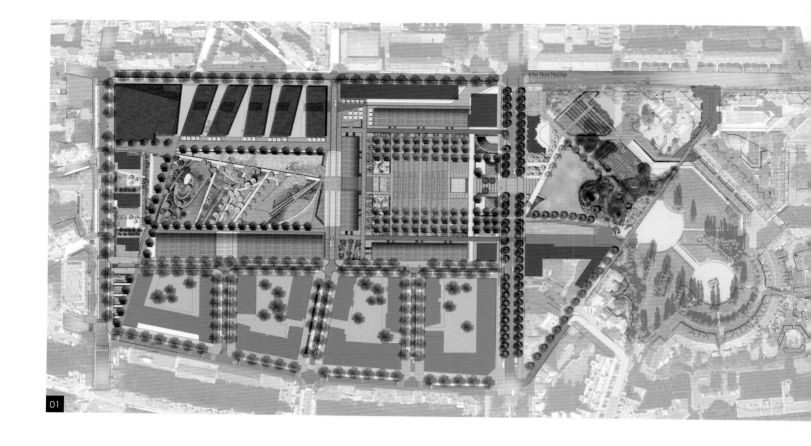

伯恩生态街区公园
Park for Bonne Eco-neighbourhood
ATELIER JACQUELINE OSTY & ASSOCIÉS

地点：法国格勒诺布尔
完工日期：2010
面积：4.5 ha
业主：SEM Sages
照片版权：AJOA (n°02-04), Claude Cieutat (n°05, 08-13)

Location: Grenoble, France
Completion date: 2010
Area: 4,5 ha
Client: SEM Sages
Photo credits: AJOA (n°02-04), Claude Cieutat (n°05, 08-13)

01. 整体平面配置图
02. 荣誉庭里的两个水池为水柱喷泉提供了框架
03. 山谷花园全景
04. 旧军营的石头被重新用到花园里来

01. Master plan
02. Dry fountains framed by two pools in the Courtyard of Honour
03. General view of the valley garden
04. Stones from the barracks reused in the garden

伯恩协议发展区公园是新建的生态小区的结构性要素。公园所在地原为军营驻地，由三个不同的空间组成。就此意义而言，公园方案与基地的过往历史产生了呼应，并呈现一个完全现代化的空间组织。

为了唤起人们对军事历史的记忆，荣誉庭花园以一个精确正方形的形式呈现：其广阔的中央空间不仅提供人们戏水活动，也可以轻松地容纳一些临时性装置，或者作为节庆集会的场所。欧什花园保留了既有的岩石堆和空地，以延续场所先前的景观风格，同时作为附近孩童的游戏场地。山谷花园与此形成对比，里头的小丘和大洼地共同创造了各种不同的景观环境与使用功能，以及位于散步小径和观景台上的众多观景点。

The park of the Bonne ZAC (comprehensive development zone) is the structuring element of the new eco-neighbourhood. Located on the site of the old military barracks, it is made up of three distinct entities. In this sense, the park echoes the site's historical past while imposing a resolutely contemporary composition.

A reminder of the site's military past, the Garden of the Courtyard of Honour forms a strict square: its large central space is taken up with playful fountains and lends itself to temporary installations and festive gatherings. With its clumps of stones and its clearing, the Hoche Garden preserves is pre-existing landscape design while attracting the neighbourhood's children to play. In contrast, the hillocks and lake of the Valley Garden combines landscape ambiances with disparate uses as well as attractive viewpoints from the planted walks or the belvedere.

05. 地面与建筑立面的构图组织
06. 山高谷花园剖面图
07. 荣誉庭剖面图
08. 水充满了荣誉庭里的水池

05. Graphic motifs on the ground and the facades
06. Cross-section of the valley garden
07. Cross-section of the Courtyard of Honour
08. Water flooding the pool of the Courtyard of Honour

然而，这三个花园组成了一个整体的公园，其东西向的主要交通轴线让人非常明显地感受到它的连续性。这条轴线与各种小路相互交错，同时也通过位于花园和城市空间之间的小广场，确保了公园与街区之间的相互渗透。除了基本元素的重复使用（水元素和植物带），被拆除的军事建筑所遗留下的石块也特别被重新利用（石梁、壁柱、花岗岩铺地），不仅重新拾起了人们对这块土地的零落记忆，也为这个多元化的公园建立起了整体性。

本方案和格勒诺布尔市获得2009年生态街区国家大奖。

These three gardens nevertheless make up a single park, whose continuity is clearly visible thanks to the line of a major circulation axis from east to west. The park's accessibility to the neighbourhood is assured by several pathways that break up this axis, as well as a system of staples and small squares between the garden entities and the urban space. Aside from the repetition of common principles (the constant presence of water, plants in rows), it is above all the re-use of the stones of the demolished military buildings (lintels, pilasters, granite paving stones) that composes a shared and disseminated memory of this unity through diversity.

This project, with the City of Grenoble, won the Grand Prix National ÉcoQuartier 2009.

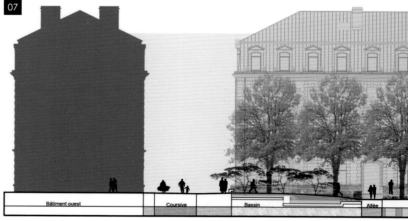

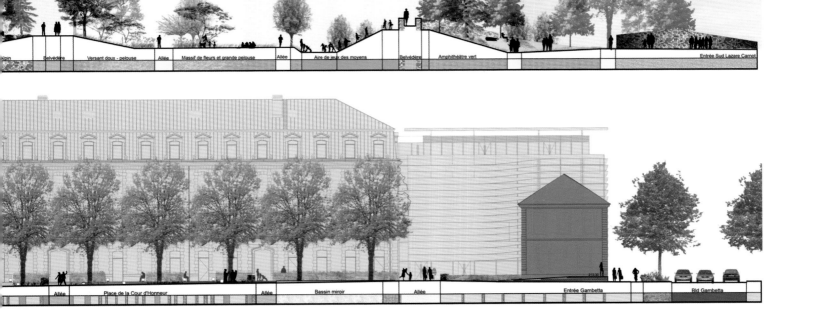

02 公园与公共性花园 Public parks and gardens

09. 儿童游戏的趣味小丘
10. 趣味小丘回应了邻近山地的地形
11. 水柱喷泉与水生植物
12. 植被细部
13. 从旧军营回收再利用的石头

09. Mountaineering in the adventure playground
10. The playground mountain reflects the surrounding scenery
11. Fountains and aquatic vegetation
12. Detail of the vegetation
13. Stones reclaimed from the barracks

02 公园与公共性花园 Public parks and gardens

桑特里居民花园
Saintryiens Garden
PHYTORESTORE / THIERRY JACQUET

地点：法国塞纳河畔桑特里	Location: Saintry-sur-Seine, France
完工日期：2011	Completion date: 2011
面积：1.5 ha	Area: 1.5 ha
业主：塞纳河畔桑特里镇政府	Client: Saintry-sur-Seine Town Council
合作设计师：Sylvestre Lieutier	Co-projet manager: Sylvestre Lieutier
照片版权：Thierry Jacquet / Phytorestore	Photo credits: Thierry Jacquet / Phytorestore

01. 方案透视图
02. 三个草原区
03. 蜜蜂草原里的小径

01. Montage image of the project
02. The three meadows
03. Path through the bee meadow

这个位于巴黎外围的小型公园项目提供了设置多样化生物环境的机会。此地块原本是埃松省列为保护的敏感自然区。项目的目标在于建设一个开放的公共空间，来说明在城市里维持多样化生态环境的重要性，因而使得一块原来不对大众开放的空间得以获得重新整治。

基地中心整治的基本原则在于开设一条连接八个不同空间的探索小径。首先是教学性的生物小区，以一个个独立的林中空地的形式展现，四周围绕着小树林和灌木丛以强化探索的氛围。在中心地块的周边设置了四个与城市连结的场所。一条教学园林大道通往一系列既存的小径、一个停车场、一个游戏场以及一个休憩空间。一个较为偏远、带有水源的地块被重新整治为天然池塘，展示着湿地的功能运作。

其他七个生物小区展现着城市与城郊的典型生态环境，可以是一道干燥石墙、一个小型蜜蜂草原、一个鸟禽栖息的林中空地…… 这里成为供学校使用的教学公园，随后再由学童们引领家长来认识此场所。此地设置的果园也令人回想起昔日具有生态作用的景观。

The project for a small public park on the outskirts of Paris was a chance to design a setting for enhancing the biodiversity of the site. Originally the plot was an Endangered Natural Space protected by the Essonne administrative department. The approach consisted of creating an open public space explaining the importance of maintaining diversified ecological habitats in the city. This project thus allowed a space that was previously closed to the public to be laid out in a new way.

The basic principle for the laying out of the heart of the site was a discovery pathway through the main redesigned plot. This pathway links eight spaces. First of all there are the educational biotopes, designed as isolated clearings protected by copses and hedges to strengthen the impression of discovery. On the periphery of the central plot are four distinct places that serve as links with the city. An arboretum walk creates a link with a network of existing paths, a car park, a playground and a relaxation space. A plot set apart, containing a spring, has been completely redesigned as a natural pool explaining how wetland zones function.

The seven other biotopes illustrate the typical habitats of urban and peri-urban environments, whether a dry stone wall, a small pollen-rich meadow, or a clearing for birds... The commune wanted an educational park that would be available for its schools, the idea being that the children would then guide their parents. The orchard also harks back to the landscapes of the past that had strong ecological functions.

04. 昆虫水池
05. 水池畔
06. 潮湿草原里的小径
07. 水池和池畔整治剖面图
08. 干石区生物群落环境剖面图
09. 干石区生物群落环境

04. A lake for insects
05. Banks of the lake
06. Path through the wetland meadow
07. Section of the banks of the lake
08. Section of the stone biotope
09. Dry stone biotope

Saules tressés

Eco-bosquet

Evocation d'un ancien mur

Evocation d'un ancien mur

02 公园与公共性花园 Public parks and gardens

10. 枯树干林中空地
11. 鸟禽小广场
12. 鸟禽小屋
13. 儿童游戏区

10. Dead tree clearing
11. Small squares for birds
12. Birds' nesting boxes
13. Children's playground

02 公园与公共性花园 Public parks and gardens

01

克雷泰伊湖岬角 *Créteil Lake Point*

PÉNA & PÉÑA PAYSAGISTES

168

地点：法国克雷泰伊
完工日期：2006-2010
面积：4.13 ha
业主：SEMIC
照片版权：Christine & Michel Péna

Location: Créteil, France
Completion date: 2006-2010
Area: 4.13 ha
Client: SEMIC
Photo credits: Christine & Michel Péna

01. 小丘沿着进入克雷泰伊市的大道而设置
02. 由考登钢堰板所支撑的一层层小平台
03. 木板栈道穿越湖畔花园而通往湖泊
01. The hill runs alongside the main approach road to Créteil
02. Small terraces held in place by Corten fascines
03. Pontoon crossing the banks garden in the direction of the lake

本景观项目与克雷泰伊湖南部新街区的建造紧密相关。此湖泊是一个旧采石场的遗迹，而新街区则沿着湖畔展开：面对一个完全天然的河岸（休闲娱乐场地）设置了一完全城市化的河岸。这个城市与自然相接的特点，为此大型景观带来独特的创意和个性。

一个景观规范在建筑物建盖之前便已经拟订出来，以要求建造商在他们各自的建地之内仍然必须对基地的大自然环境予以尊重并纳入设计考量，景观的影响评估不应当仅止于地块的边缘界线。方案设置了一系列连接着各个地块并缓缓下降到湖面的平台，建筑物则建盖在这些边缘以石笼挡土墙支撑的平台上。每个地块下方的边界以与湖泊曲线相应的线条来界定，而非依照初始计划来建造2米高的冷峻围篱。这个边界曲线随着高度的差异而形成了同时是围墙也是阳台的空间。建造商同意将位于曲线以外的土地归还给市政府，此处设置的一个大型湿地植物园将公众空间与私人空间隔离开来，一条木造浮桥穿越其间让人得以在花园中散步。根据不同基地高度而形成的景观层次同样运用在街道行道树的选择上：高处种植的树种与湖畔的树种是有所区别的。

This project is linked to the creation of a new neighbourhood south of Créteil's lake. The town was built up around this lake, formed from an old gravel pit, in a very landscaped way: a bank that is entirely urbanised faces a bank that is entirely natural (the leisure park). This town-nature cross formed the originality and character of this important landscape.

A landscape charter was put in place before construction began on the buildings to ensure that the developers took into account and respected the qualities of the site even from inside their plots, as the landscape does not end where land ownership begins. Terraces descending to the lake were proposed, running along the plots and sustained by gabion walls on which the buildings would be placed. At the lower limits of the plots, rather than the austere 2 metres-high fence that was originally proposed, a large curve has been drawn following that of the lake. Making the most of the difference in levels, it forms a large balcony wall. The developers agreed to give back to the town the land situated beyond this line. A large garden of wetland plants was created, keeping a distance between the public areas and the private gardens. A wooden pontoon runs around it. The landscape stratification of the site continues in the planting of rows of trees along the roads: the species on the high part are not the same as those close to the lake.

04. 沿着围墙阳台边缘设置的湖畔花园保护着住宅的私密性
05. 位于公共空间与私人住宅之间的花园
06. 在高处设置木质铺板让人得以眺望湖泊景致
07. 从丘陵往下的阶梯

04. The banks garden beside the balcony wall that separates the private residences
05. The wetland garden in the space between the public park and the residences
06. High decking gives a view over the lake
07. Steps down the hillside

此基地原本是垃圾倾倒场，因此必须进行重要的去污染工程。大量的废石土方经过筛选之后必须在基地附近被重新使用。在道路设施与新街区之间不可建造的地块上，运用填土的围积与塑型形成了一些大型的观景台，以供人们欣赏城市与湖泊全景。以考登钢制成的堰板得以在斜坡上支撑土壤来种植大量的植物。配合南向坡面的高温干燥，这里种植了一系列地中海植物，它们自幼株即栽植于此，以便在成长过程中更完美地适应基地的形态与气候。

As the site was originally landfill, a major clean-up programme had to take place. Once screened, large amounts of rubble had to be reused in the immediate surroundings of the site. Large lookout towers were built, giving a view of the whole town and the lake and finding a use for the rubble and the contours of the non-buildable land that situated between the infrastructure and the new neighbourhood. Corten fascines provided a way of holding the planting earth in place and creating substantial plantations. On the south side, making the most of this very warm and dry situation, a Mediterranean palette has been composed, using young plants in order to guarantee a perfect adaptation.

02 公园与公共性花园 Public parks and gardens

08. 以考登钢制成的堰板
09. 这里种植着地中海植物，主要为鼠尾草、石蔷薇和橄榄树
10. 以考登堰板迎接南向坡面的高温干燥
11. 小丘上的橄榄树生长很好

08. Corten fascines
09. Mediterranean beds with planting based on sage, cistus and olive trees
10. Corten fascines on the south side
11. Olives grow well on the hillside

02 公园与公共性花园 Public parks and gardens

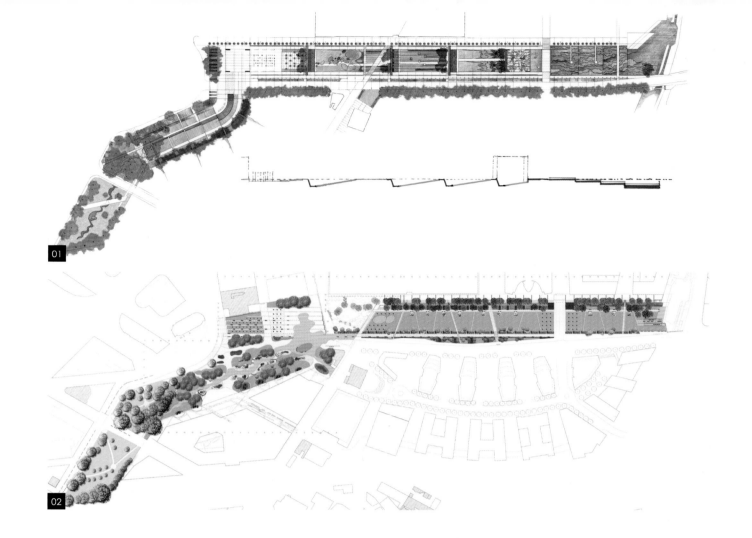

东区城市公园
Eastside City Park

ALLAIN PROVOST & PATEL TAYLOR

地点：英国伯明翰
完工日期：2012
面积：3.2 ha
业主：伯明翰市政府
照片版权：Tim Soar

Location: Birmingham, United Kingdom
Completion date: 2012
Area: 3.2 ha
Client: Birmingham City Council
Photo credits: Tim Soar

01. 2006年设计竞赛获选的平面图
02. 2006年申请建造的平面图
03. 公园鸟瞰透视图和街区新发展
04. 水渠景观透视图
05. 从北部望向照明结构体和阶梯式平台

01. Landscape competition plan 2006
02. Landscape planning application plan 2009
03. Aerial view of the park and the new development
04. View along the canal promenade
05. View towards the lighting fins and terraces from the north

伯明翰位于几个卫星城市零散结聚成的大区域（约400万人口）的中心，带着19与20世纪工业变迁所留下的城市混乱现象。在东区建设一个长线型的公园成为城市更新的重要一步，此项目有如在难以辨识的城市中心地带展开一场手术。

方案最初构想在于以一个几何构图来与基地的混乱背景作为抗衡，创造一个合理而清晰可辨的架构，其上设置各式花园，仿佛一个在600米长空间上展开的绿色折纸艺术。在公园的两端原本计划设置门廊来界定边缘，像个充满节庆氛围的荧屏。在公园的长向边缘设置两条小径，北侧小径沿路种植被裁剪为金字塔形的千金榆，南侧小径则有椴树遮阴。这两条南北边缘的小径同时也作为高起的散步平台，让人得以居高观赏公园中央由主题花园所组成的长条绿毯，此绿毯被切割为一个个斜坡，彼此之间以小瀑布相连接，几条紧邻绿毯的低处小径可以带领人们进入其中探索。

自从2005年设计竞赛完成之后，经济压力迫使设计师在遵循原方案基本精神的前提下，对设计进行大幅修改以简化方案。折纸艺术的构想被一系列将基地进行几何分割的线条所取代，由植物、河渠和小径共同构成一个明显的空间架构。在此架构上随着植物的变化而添加了诸多细部处理，以便为每个场所带来独特的氛围。不幸的是，位于公园边上的博物馆最后强制要求在公园内设置一个科学花园，此花园不仅无法融入整体的结构文脉，还将公园一切为二，于是逆转了公园的故事⋯⋯

Birmingham is the centre of numerous satellite cities, with a population of 4 million. The urban chaos was shaped by 19th- and 20th-century industrialisation. The regeneration of the central Eastside district was begun by a linear park that links the city centre to a formerly industrial area.

The initial concept involved the introduction of a strong geometry, in contrast to the randomness of its context: a legible frame, coherence, and varied planting strategies within different gardens, extending over 600 metres. Lengthways, a row of pleached hornbeams to the north of the park, and a row of pleached limes to the south, frame a series of sloping thematic gardens with waterfalls and sunken paths.

Since the competition in 2006, the scheme has undergone drastic simplification due to a reduction in the budget, though the spirit of the original design has been maintained. The giant origami of the sunken gardens was replaced by a series of parallel forms: a strong frame of planting, canals and pathways which link numerous smaller, more intimate spaces, each with their own unique planting. The museum, which borders the park, has added a science garden, but the design is not integrated into the park, and, alas, the park is cut in two.

02 公园与公共性花园 Public parks and gardens

06. 从西边望向阶梯式平台
07. 灯座的植物装饰图案
08. 朝向阶梯式平台上方的景观
09. 2012年的建造平面图

06. View from the west towards the terraces
07. Detail of the leaf patterns on the lighting fin cladding panels
08. View from the top terrace past the lighting fins
09. Construction plan 2012

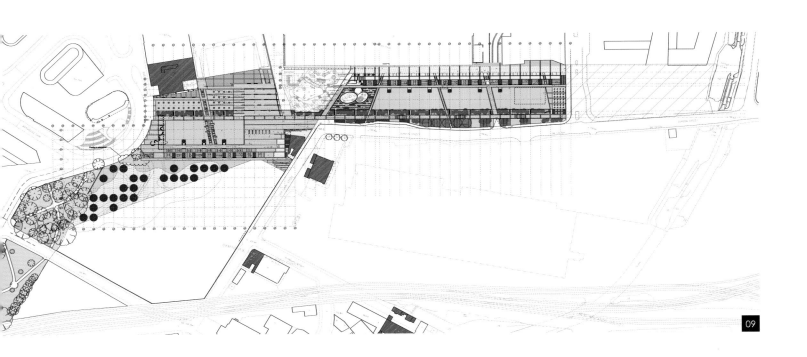

02 公园与公共性花园 Public parks and gardens

10. 顺着中央草坪望向公园
11. 从凉棚望向植物帷幕
12. 从主要散步道望向凉棚
13. 从公园望向城市中心
14. 裁剪整齐的千金榆

10. View along the formal lawn
11. View to planting screen from pergola
12. View along the spine path to the pergolas
13. View towards the city centre
14. View towards the pleached carpinus betulus

02 公园与公共性花园 Public parks and gardens

圣彼得堡动物园
Saint-Petersburg Zoo
TN PLUS

地点：俄罗斯圣彼得堡
完工日期：2014
面积：300 ha (总基地), 96 ha (动物园用地)
业主：圣彼得堡市政府
合作设计师：Agence Beckmann-N'Thépé (设计竞赛阶段)
图片版权：Artefactorylab

Location: Saint-Petersburg, Russia
Completion date: 2014
Area: 300 ha (whole site), 96 ha (zoo area)
Client: Saint-Petersburg City Council
Co-project manager: Agence Beckmann-N'Thépé (competition phase)
Image credits: Artefactorylab

01. 方案全景透视图
02. 塑造"泛古陆"的构想成为圣彼得堡动物园的设计基本准则
03. 整体平面配置图

01. Overall view
02. The concept of Pangea is the founding principle of Saint-Petersburg Zoo
03. Master plan

圣彼得堡的新动物园处于一片广阔的原始土地上，它虽然远离城市历史中心，但是并没有与世隔绝，因为它同时是云托罗维斯基的自然保护区、附近居民的散步场所以及特殊动物群和植物群的摇篮，这些都成为项目不可忽略的环境特点。为了尊重这些元素的需求，动物园方案仅设置在大约三分之一的基地面积上，而让基地内其他大范围土地能够不受外力介入而自由生长。方案从关怀环境的角度展开：通过了解生态环境的历史，以便能够有效率地保护生物多样性。圣彼得堡新动物园的构思正是诞生在这个基础上。

方案的主要构想在于将如今相互分离的生物群落做适当的整合：为每一块陆地选择一些具象征性的物种来创造一个仿佛重建"泛古陆"的幻想。重新组合的群岛展现了东南亚洲、非洲、澳洲和南美洲，以及由北极浮冰连接的北美洲和欧亚大陆。由于方案基地中存在大量的水分，这个生态岛屿组合的构思显得特别合适。

方案所规划的每个围场都是整体景观的一部分，模拟着它所代表的原始陆地的生态环境。其地形、植物覆盖、水文都给人一种身处非洲撒哈拉沙漠南部的热带草原，或者南美洲潘帕斯草原中的感觉。方案的建筑物也被融入岛屿景观规划之中，根据各自的功能，它们或伪装在环境中，或清晰可见。随着每个岛屿的特色而建造，其上的建筑都呈现出简约、独特、完全当代的风格。

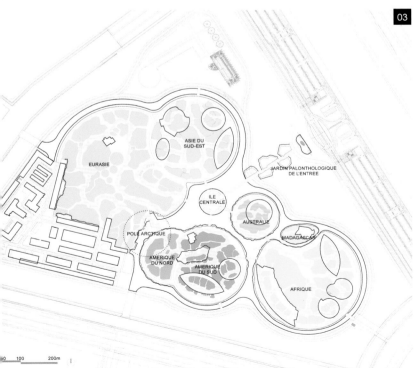

Though this immense virgin expanse is far removed from the historic centre, the site of the new Saint-Petersburg zoo is not isolated. The protected zone of the Yuntolovsky nature reserve is a place where the locals love to walk and the cradle of an exceptional fauna and flora – assets that should not be discarded lightly. The project is taking place here in a spirit of respect for these elements, on a third of the land area, leaving a large part of the site free and intact. Thus it aims to establish an approach that is concerned with its environment: the effective protection of biodiversity through the understanding of its history. On this premise that the new Saint-Petersburg zoo was born.

The main idea of the project is to bring together and compare biotopes that are today separate: for each continent a symbolic sample of species is offered in order to create the illusion of a reconstituted Pangaea. The archipelago thus composed will represent South-East Asia, Africa, Australia, South America, North America and Eurasia, the latter two being linked by the ice floes of the North Pole. Thanks to the abundant presence of water, the site is particularly well adapted for the organisation of the habitats into islands.

Each enclosure will be part of a global landscape inspired by its original environment. The topography, the plant cover, the hydrography will give the impression of being in the savannahs of Sub-Saharan Africa or in the pampas of South America. The architectural structures will be integrated into the landscape of the islands, camouflaged or clearly visible according to their function. The architecture will be specific to each island and expressed by objects with simple forms, unusual and resolutely contemporary.

04. 非洲园地
05. 北极区
06. 动物园入口

04. The Africa space
05. The North Pole
06. The zoo entrance

02 公园与公共性花园 Public parks and gardens

艾夫兰山公园　*Mount Évrin Park*

URBICUS

地点：法国蒙泰夫兰
完工日期：自2005年起施工
面积：20 ha
业主：马恩河公共整治机构、蒙泰夫兰镇政府
照片版权：Charles Delcourt (n°01, 03, 05-07, 10-12), Michel Reuss (n°04, 08, 09)

Location: Montévrain, France
Completion date: ongoing since 2005
Area: 20 ha
Client: EPA Marne, Montévrain Town Council
Photo credits: Charles Delcourt (n°01, 03, 05-07, 10-12), Michel Reuss (n°04, 08, 09)

01. 大草原为城市提供了广阔的自然场所
02. 此公园成为未来生态街区的空间脊柱

01. The large meadow offers a natural setting within the town
02. The park is the backbone of the future eco-neighbourhood

艾夫兰山公园坐落于未来蒙泰夫兰生态街区中，分布在一个将近2千米长的线形空间里，占据大约20公顷的用地。它将连接位于马恩河丘陵地的老城镇中心和位于欧洲之谷街区南部的格赛小溪谷，成为一条大城乡区域尺度的生态廊道。此公园"萃取"了当地地理和生态环境的特色景观而进行组构，以确保乡野景致的活力与多样性能够在这块土地上获得保存。

为了尊重环境与其可持续发展，艾夫兰山公园被构思为乡村公园，根据公园提供的不同活动而组织理性且灵活的维护管理，此乃所谓的差异化管理。这个具有弹性的管理方法以细心的观察为前提，以便栽种最适合环境的植被，让植物自由生长，而人为的处理只有在绝非必要时候才介入。

为了使大众接受这种新的维护方式，必须借助设置教学性的标示牌，以提高人们对环境的敏感度和认知理解。

At the heart of the future eco-neighbourhood of Montévrain, the park of Mont Évrin will eventually occupy an area of around 20 hectares on a strip of around two kilometres. It will link the old town of the commune on the hillsides of the Marne with the small valley of the Gassets brook in the south, situated in the Val d'Europe neighbourhood. It will become an important ecological corridor on the scale of the urban area. Made up of "extracts" of landscape characteristic of the surrounding geography and habitats, the park aims to preserve the dynamic and diversity of the rural landscapes of this land area.

In a spirit of respect for the environment and sustainable development, the park of Mont Évrin has been designed as a rustic park, calling for a well thought-out management strategy that can be modified according to the park's use, which is known as differentiated management. This fluid approach allows us to follow and refine the installation of the vegetation by observing in order to prioritise the best adapted planting, and not to intervene except when truly necessary.

The public's acceptance of this new form of management called for an awareness campaign, which was undertaken in the form of educational panels.

03. 用树木围隔的草地
04. 大果园里的通道之一以及其路边的草沟,夏天绿意盎然的景观
05. 草原的管理方式使得空间能够随着季节而弹性变化
06. 这个位于一个住宅街区中央的公园提供给居民一个方便使用的休闲空间
07. 休憩空间融入绿化空间之中

03. Traditional hedges and small fields landscape
04. One of the pathways through Grand Orchard and the swale in summer
05. The management strategy for the meadows allows for shaping the spaces according to the seasons
06. The park, at the heart of a residential neighbourhood, offers spaces that are easy to use
07. Resting places are created in the heart of the vegetation

02 公园与公共性花园 Public parks and gardens

08. 春季繁茂的植物使得建筑物完全改观
09. 私人住宅区的入口设计采用着公园的植物语汇
10-12. 公园中的穿越小径和道路小品以条状木板与镀锌钢展现出最和谐的搭配

08. In spring, the vegetation transforms the look of the buildings
09. Private entrances have been designed using the vocabulary of the park
10-12. Furniture and pathways combine wooden decking and galvanised steel

02 公园与公共性花园 Public parks and gardens

03

Urban Public Spaces
城市公共空间

01

保罗·格里莫花园广场
Paul Grimault Esplanade-Garden
AGENCE APS

地点：法国安纳西
完工日期：2011
面积：3 400 m²
业主：安纳西市政府
照片版权：Pierre Vallet (n°02-06), Romain Blanchi (n°07-10)

Location: Annecy, France
Completion date: 2011
Area: 3 400 m²
Client: Annecy City Council
Photo credits: Pierre Vallet (n°02-06), Romain Blanchi (n°07-10)

01. 平面配置图：一个为城市进行"缝合"的项目
02. 花园广场重新建立起与历史性城市的关系
03. 城市平台上的日光浴场散发出闲散的氛围
01. Plan for a project defined as "urban couture"
02. The esplanade-jardin re-establishes the links with the historic city
03. The sun-bathing deck of the urban terrace invites relaxation

保罗·格里莫花园广场介于库里耶协议开发区的新街区和安纳西城市的历史街区之间。在改善不受欢迎的"楼板平台式城市规划"（1960年代法国主张人车分离的城市建设方式）这个目标的驱使下，这个项目通过对基地的提升转化和促进使用而赋予它新的活力。APS事务所在此营造了便于重塑历史的诗意氛围和能够适应项目所建构之新世界的空间语汇，透过三个明显区别的空间各自的尺度和特质来重新划分这个空旷的大型基地。

北侧，三角形的"绿色广场"建立了项目与布洛尼林荫道的关系。其平坦的场地和硬质铺地的独特图案反映出帕尔默兰山区石灰岩沟的特色，这是安纳西附近地理和地质方面的标志性特征。这个改造"楼板平台式城市规划"的项目重引进了"城市中的大自然"这个主要概念，以提升地面的生命力。波浪形状的长条裂缝展现在铺地之中，并令人难以置信地看到了禾本植物和多年生植物从中渗透出来。

一片长方形城市广场坐落于高处平台的南侧、面向附近的群山，成为这个崭新公共空间的心脏。其木质铺地上设置了由考登钢制成的圆锥形花盆，展示着成群的樱桃木；扶手椅和长椅则展现出日光浴场的慵懒氛围。在平台西侧和东侧边缘的两个花园中，考登钢板和繁茂的多年生植被相互交错成"钢琴琴键"式的图案。"玫瑰花园"的坡面享受着其向南的绝佳处境，把一系列不同品种的古老玫瑰都展现出来，令人想起阿尔卑斯山区的那些野玫瑰。

The proposal for the Paul Grimaud esplanade-garden aims to reactivate a site by transcending it, effacing the unwelcoming 1960s-era functionalist urban planning known as "Urbanisme de Dalle". It is designed to encourage its appropriation by the inhabitants as well as to form an interface between the new neighbourhood of the Courier ZAC (comprehensive development zone) and the historic centre of Annecy. The agency APS has designed a poetic of situation to reforge a history and rhetoric appropriate for the new, rebuilt world of the project. Three distinct entities will give the large vacant space a new dimension through their scale and their distinct natures.

In the north, the triangle of the "planted esplanade" forms a wedge of garden next to Brogny Avenue. The use of planes and the unusual vocabulary of the hard surface plaza recall the limestone formation of the Parmelan lapies, an emblematic geographical and geological feature near to Annecy. The project for transforming the concrete environment brings back the federating idea of nature in the city to reactivate the fertility of the soil. In the new treatment of the plaza undulating faults mark out a surprising pattern of strips of grasses and perennials.

Elevated on the south border of the high plateau, the urban terrace, a large rectangle with views of the neighbouring mountains, becomes the beating heart of the new public space. On the composite wooden decking, clumps of flowering cherries set the scene in their colourful cones of Corten steel. Chairs and loungers give a languid feel to the sun-bathing area. At the eastern and western ends of the terrace, the two "piano keys" gardens alternate strips of Corten with generously planted beds of perennials. The "rose garden" makes the most of its south-facing situation on a slope to display a collection of old rose varieties, evoking the wild roses of the Alps.

04. 玫瑰花园和城市平台
05. 方案的三个各具特色的空间展现在此鸟瞰景致上
06. 位于平台两端的两个地面构图犹如"钢琴琴键"的花园

04. The rose garden and the urban terrace
05. A trio of atmospheres work together
06. Two "piano keys" gardens at the ends of the terrace

03 城市公共空间 Urban public spaces

07. 广场上的"花园化大自然"是附近山区景致的影射
08. 此花园广场方案使"楼板平台式城市规划"获得新生
09. 禾本植物和多年生植物从波浪形状的长条裂缝中渗透出来
10. 禾本植物、多年生植物和水泥铺板

07. The "managed nature" of the esplanade echoes the surrounding mountains
08. The esplanade-garden transcends "urbanisme de dalle" (a form of urban planning popular in France in the 1960s where pesdestrians areas "over structure" were seperated from vehicle traffic)
09. Undulating strips of grasses and perennials
10. Grasses, perennials and concrete slabs that look like natural stone

03 城市公共空间 Urban public spaces

"Inhabited Wood" – Euralille 2
"林中居所"–欧洲里尔计划2
AGENCE TER

地点：法国里尔
完工日期：2012
面积：15 ha
业主：欧洲里尔混合经济开发公司
合作设计师：MG-AU architectes urbanistes, Dusapin-Leclercq architectes
照片版权：Camilla Pongiglione / Agence TER (n°03-05, 07), Giovanni Nardelli / Agence TER (n°06)

Location: Lille, France
Completion date: 2012
Area: 15 ha
Client: SAEM Euralille
Co-project manager: MG-AU architectes urbanistes, Dusapin-Leclercq architectes
Photo credits: Camilla Pongiglione / Agence TER (n°03-05, 07), Giovanni Nardelli / Agence TER (n°06)

01. 整体平面配置图
02. 街区三维效果图
03. 绿化环境中的树木网格
04. 植被效果消弭了私人空间的界线

01. Master plan
02. 3D plan of the neighbourhood
03. Grid of trees on the green
04. The planting blurs the boundaries of the private areas

"林中居所"这个名称一下子就揭露了街区的主要规划原则：绿化在横向和竖向维度上向街区的所有街道铺展。这片林中居所的中心构想在于公共空间的设计，由于停车空间的地下化，使行人得以尽情享用室外空间，与土地保持恒常的联系。

景观师建议在施工之前预备好350棵大型树木（美国皂荚树、日本槐树和木兰），以便在施工过程陆续种植。位于街区内部植被丰富的中央庭院成为东西向住宅私人庭院的翻版，前后两个庭院改变了住宅临街和临内院的传统布局形式。水的管理借助住宅庭院中的防洪草沟完成，这些位于住宅楼和公共空间之间的空隙收集着居民的用水，在合适的时机过滤后加以利用，多余的则通过排水系统排走。

被完全隔离在街区之外的汽车仅能通过外围环路进入半地下停车场。"林中居所"由从前的飞地转变为一个真实的避风港，那些高出第一批建筑物的树林已经成为这个事实的见证。

The name "Inhabited Wood" immediately reveals the founding principle of the neighbourhood: it is plunged into the vegetation that extends both horizontally and vertically and is a part of every street in the neighbourhood. At the heart of this inhabited wood is the use of underground parking so that the pedestrian finds himself master of the overground world, in a constant relationship with the earth.

The landscape architects recommended that 350 large trees (American honey-locust trees, Japanese sophoras and magnolias) be reserved even before construction work began, in order to be planted as the building site progressed. A central tree-planted green mirroring the private gardens with an east-west aspect replaces the traditional typology of street side/garden side. Water management is via swales in front of the houses. These ditches between the buildings and the public space harvest rainwater from the houses to temporise and percolate it, the surplus being evacuated via a drainage system.

Banished from the neighbourhood, the car is consigned to a peripheral ring-road which feeds into the half-underground car parks. Formerly hemmed in, the "Inhabited Wood" has now become a veritable haven, as the treetops that rise above the first constructions already testify.

05. 从街道到建筑物脚底的景观处理
06. 绿化空间成为建筑与街道之间的过渡
07. 受到绿荫遮护的中央庭院散布着可供休憩的椅凳
08. 标准段落剖面图：从住宅楼上眺望中央庭院和绿化斜钩
09. 基地横向剖面图：地下化的停车场有利于公共空间的设置

05. Landscape treatment at the foot of a building on the street side
06. The vegetation creates the transition between the buildings and the street
07. A central shady green with scattered benches
08. Section: the landscaped swales and the central green seen from the upper floors
09. Cross-section: car parking is placed underground to improve the public space

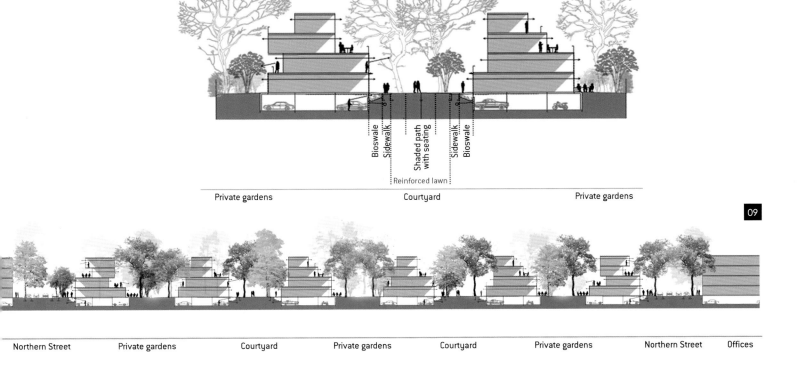

201

03 城市公共空间 Urban public spaces

城堡广场 *Schlossplatz*

AGENCE TER

地点：德国卡尔斯鲁厄
完工日期：2011
面积：18 500 m²
业主：巴登-符腾堡州政府房地产与建筑部门
照片版权：Atelier Altenkirch

Location: Karlsruhe, Germany
Completion date: 2011
Area: 18 500 m²
Client: Vermögen und Bau Baden-Württemberg Council
Photo credits: Atelier Altenkirch

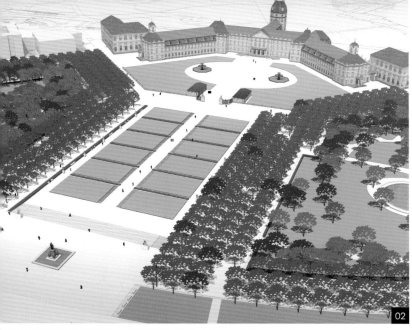

01. "凯旋大道"是连接城堡与城市的轴线
02. 鸟瞰透视图，整修后的新阶梯和园圃
03、04. 水渠线条直指着园中的雕塑

01. The "Via Triumphalis" is an axis linking the castle to the town
02. Aerial drawing of the new staircase and the reworked parterres
03-04. The linear pools reflect the sculptures

几个世纪以来，位于城堡和城市之间的城堡广场不断地经历改造，使得要想纯粹地还原某一段历史已成为不可能。与其在不同历史阶段中选取一个并将其形式化，岱合景观事务所以最显著的历史印记为基础，把这些在今天看起来仍然合理且让人接受的形貌以当代的手法进行诠释。

朝向略微下沉的绿坛区的一片广阔空间和占据了两侧小灌木丛之间所有长度的阶梯组成了城市和城堡广场之间的过渡场所。在此之前，人们很难发现隐藏在灌木丛中的雕像，而现在它们都被刻意强调出来，围绕着中央绿坛而排置。边缘以金属围合的绿坛稍稍高于地面并铺设着草皮，把古典型体转换成现代语汇；长条状的水池在草地间划出横向的条纹，一直延伸至雕像脚下，映照出雕像和树木繁茂枝叶的倒影。

绿坛和小灌木丛之间的高差被由矮墙和树篱所组成的宽厚边界所吸收，树篱中安置着各种设施。绿坛内部和主楼梯周围的照明皆非常简约且贴近地面，而矗立在树木间的灯柱则照亮着小灌木丛。

Situated between the castle and the town, Schlossplatz was constantly remodelled over the centuries, making a purely historical approach impossible. Rather than choosing one of these different periods and designs, Agence Ter proposed a contemporary interpretation based on the most significant historical aspects that still seem appropriate and acceptable today.

A wide open space forms a threshold between the town and the castle square. It descends towards the slightly sunken parterre via a wide staircase running the whole width between the two groves. The sculptures, which had been hidden inside the groves, are made more visible and frame the central parterre. The latter, transposing a classical motif via a contemporary vocabulary, is composed of slightly raised lawns framed by metallic borders; it is striped crossways by thin lines of water leading to the sculptures, which are reflected in them, along with the foliage of the trees.

The change in level between the parterre and the groves is marked by a thick border made up of a wall and a hedge incorporating furniture. The lighting is minimalist and horizontal in the parterre or around the main staircase, while lamp-posts rise between the trees to light up the groves.

05. 主要轴线道路以地面灯光强化出个性
06、07. 通过地面灯光的照射，彰显出园中的历史性雕塑
08. 水渠与绿坛的相遇
09. 从阴凉的树丛到艳阳下的绿坛
10. 水渠为绿坛画上一条条横纹
11. 水是吸引人且充满趣味的元素

05. Ground-level lighting enhances the main avenue of the park
06-07. Atmospheric lighting for the historic statues is implanted in the ground
08. The linear pools meet the flower beds
09. From shady groves to sunny parterres
10. The parterre striped by lines of water
11. Water is both attractive and fun

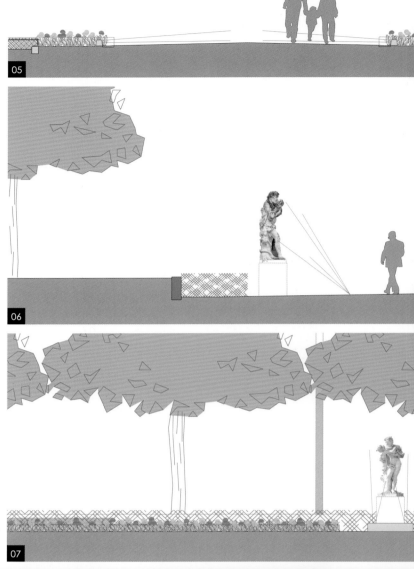

03 城市公共空间 Urban public spaces

夸斯兰停车场
Coislin Car Park
ARTE CHARPENTIER

地点：法国梅斯
完工日期：2011
面积：1 ha
业主：Q Park
图片版权：Géraldine Bruneel (n°02, 03, 05-07), Élodie Ledru (n°01, 04)

Location: Metz, France
Completion date: 2011
Area: 1 ha
Client: Q Park
Image credits: Géraldine Bruneel (n°02, 03, 05-07), Élodie Ledru (n°01, 04)

01. 整体透视图
02. 图案效果
03. 绿色沙龙为停车场带来韵律感
01. Overall view
02. Graphic motifs
03. Green rooms break up the uniformity of the car park

本项目通过一个高质量的整治方案来重新构建广场上的停车场。此广场位于靠近市中心的一个设施齐全的街区，其上穿梭着停车场的使用者、自行车和行人。两个关键点引领着夏邦杰建筑设计事务所的景观师的思路：一方面，重新赋予此停车场高度的城市化质量；另一方面，在保护措施的限制下为广场找到绿化的方法，因为这个广场必须预留将来进行考古挖掘的可能性，因此植物种植不能开挖超过30厘米的深度。

为了达到目标，景观师进行了两个策略措施。首先是创造一些"绿色沙龙"，几组大花池占据了停车场的两个车位，每组都由三个不同高度的的立方体构成，配套的长凳布置在临通道的一侧。立方体花池中种植着经过挑选、枝干挺拔的树种，这些绿色沙龙构成造型各异的小树林，自由地分散在停车场中。它们所形成的植物丛在视觉上打断了停车场内汽车的单调排列，比单独种植的树木产生更显著的效果。在每个沙龙中也设置了两个高度不同、形式简洁的照明灯，如"花束"一般竖立其中。

The project involved restructuring the car park on this square in order to improve the quality of the environment. Situated in a neighbourhood well provided with facilities, near to the town centre, the square is used by car owners, cyclists and pedestrians. Two aims guided Arte-Charpentier's thinking. On the one hand, to improve the quality of the car park as a urban facility, and on the other to find a way to plant it despite a preservation order that forbids planting deeper than 30 cm in order to preserve the possibility of a future archeological dig.

Two strategies have been employed. This first consisted of creating "green rooms", effectively large garden boxes occupying two parking spaces. They are composed of three cubic volumes of different heights, together with benches for the rooms situated along the paths. Planted with a selection of tall standard trees and shrubs, these green rooms form clumped groves spread randomly about the car park. In this way plant masses are obtained that visually cut the rows of cars and offer a more marked plant presence than with isolated trees. Lighting "bouquets" have also been installed in these rooms in twos, using minimalist lamp posts at different heights.

04. 绿色沙龙模组1和模组2
05. 在步行道一侧设有长凳的一组绿色沙龙
06. 自行车停放处
07. 小广场沙龙

04. Green room modules type 1 and 2
05. A green room with a bench running alongside the pedestrian path
06. The cycle shelter and bike boxes
07. The square's green room

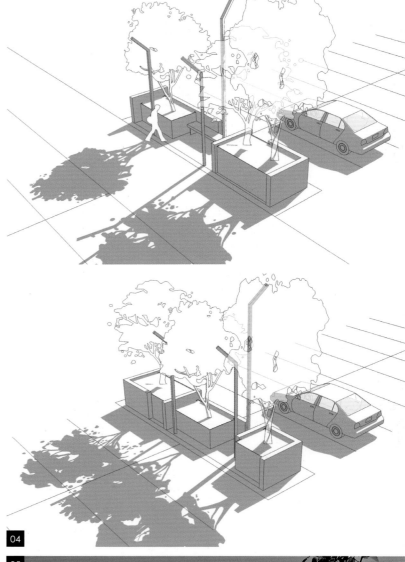

另一个策略措施的实现充分利用了一个旧地下停车场，把它拆除后填入植土以便在没有深度限制的土地上种植植物。这个旧地下停车场所提供的可能性被大幅度利用，因而得以塑造一片绿化广场，这也为周围居民提供了一片退缩于街道的休闲场所。此处的植物栽植仿照果园的风格，其中的果树按照4米x4米的网格种植。管理办公室、停车场收费站以及自行车停车位都布置在广场侧面一栋线条简洁的建筑中，此建筑采用了与绿色沙龙同样的材质。因此停车场犹如一个广场，呈现出另一种气质与风情。

The second strategy was to expropriate an old underground car park that could be demolished to restore the possibility of planting in real earth. This approach enabled the creation of a square in tune with the life of the inhabitants, offering a place of relaxation set back from the street. The planting is in the spirit of an orchard, composed of fruit trees planted in a four by four metre grid. The management office, the parking payment meters and the cycle parking take their place along the square, in a building with pure lines in the same materials as the green rooms. As a result the car park now has more of the attractiveness associated with a square.

03 城市公共空间 Urban public spaces

拉马丁广场
Lamartine Esplanade
ATELIER VILLES & PAYSAGES

地点：法国马孔
完工日期：2008
面积：3 ha
业主：马孔镇政府
合作设计师：Garcia-Diaz (设计总负责), Ingerop
照片版权：L'Atelier Lumière (n°05, 08, 09), Balloïde-photo (n°01, 03, 04), Jean-Paul Planchon (n°06, 07), Patrick Georget / Ville de Mâcon (n°02)

Location: Mâcon, France
Completion date: 2008
Area: 3 ha
Client: Mâcon Town Council
Co-project manager: Garcia-Diaz (project representive), Ingerop
Photo credits: L'Atelier Lumière (n°05, 08, 09), Balloïde-photo (n°01, 03, 04), Jean-Paul Planchon (n°06, 07), Patrick Georget / Mâcon Town Council (n°02)

01. 方案全景
02. 浪漫花园展现出一系列独特植物
03. 方案的结构线条平行于6号国道和索恩河

01. Overall view of the project
02. The romantic garden is composed of native species
03. The construction lines of the project follow the parallel lines of the RN6 highway and the Saône

拉马丁广场是一个位于索恩河及6号国道之间占地3万平方米的广大空间，其整治项目的目的在于为老城市中心和索恩河岸之间建立起对话关系。方案的基本构思在于建立一个"城市基座"，以和谐的处理方式来使广场与市政府的建筑立面达成协调。

这个整治项目首先将机动车从空间中撤离，这里最初被一片露天停车场所占据，在改造后则成为焕然一新的广场，同时建设的三层地下停车库提供了300个车位。面向河流的新设施为公共空间带来活力：一个漂浮的舞台和其观众台阶，以及几个提供小餐点的亭子。索恩河岸的整体改造旨在建立一条连续的沿河步道，它也是"蓝色道路"发展规划中的一部分。

Lamartine Esplanade forms a vast triangular space of 30,000 m² between the Saône and the RN6 highway. The challenge of its development was to forge a new urban dialogue between the old town and the Saône banks. The project was founded on the creation of a "town plinth" that used a coherent treatment to connect the esplanade to the facades of the town hall.

First step in the development was to free the space from cars. Initially occupied by a surface car park, the esplanade has been entirely redesigned and an underground car park on three levels today offers a capacity of 300 parking spaces. New facilities connected to the Saône have enlivened the public space: a floating stage and its terraces, as well as snack bars. The quays in their entirety have been redeveloped to form a new pedestrian continuity which forms part of the development plan for the "blue way".

04. 梧桐树林荫道下设置了一个停车场，并且容纳着每周一次的露天市场
05. 双杆路灯强化出一旁经过的6号国道
06. 特殊的河岸舞台提供了一个面向河流的表演场所
04. The avenue of plane trees hosts the weekly market and local parking
05. Double-column lamp posts line route of the RN6
06. The floating stage offers a place for performances on the river

此方案以简洁、婉约的形式以及和谐的色调来创造一个具有现代气氛的新场所，与城市历史交融在一起。在此，我们设计了三个朝向索恩河的新公共空间：利用现有的梧桐树林荫道来设置便利的停车位和周末集市；在硬质铺地的广场平台旁设置错落有致的台阶将人们引向索恩河景观；透过浪漫花园以及不同主题的花圃来对拉马丁这位出生于马孔的杰出诗人的致敬。从他的诗作中摘录出的句子被刻在铺地的花岗石上，点缀着整条人行步道。

The project prioritises simple and subtle forms and harmonious tones to create a new place with a contemporary appearance, which nevertheless harmonises with the history of the town. It offers three new public spaces connected with the water: the existing grand avenue of plane trees, which hosts local parking and the weekly market; the hard surface esplanade that descends towards the Saône through a series of steps; and the Romantic Gardens organised around floral themes in homage to Mâcon's illustrious native poet.

03 城市公共空间 Urban public spaces

07. 圣罗兰桥为方案的空间组织画下句点
08. 当夜幕降临，广场的地面出现灯光投影，为空间带来活力
09. "露天阅读室"配备着具有现代感的座椅

07. The Saint-Laurent bridge completes the composition
08. At night the square is enlivened by light projections on the paving stones
09. An open-air reading room with contemporary urban furniture

03 城市公共空间 Urban public spaces

Zénith Concert Hall Public Realm
天顶音乐厅外围空间
ATELIER VILLES & PAYSAGES

地点：法国斯特拉斯堡
完工日期：2009
面积：30 ha
业主：斯特拉斯堡城市委员会
合作设计师：Egis France, Bécard & Palay
照片版权：Atelier Villes & Paysages (n°02, 16), Balloïde-Photo (n°01, 08-11, 14, 15), L'Atelier Lumière (n°03)

Location: Strasbourg, France
Completion date: 2009
Area: 30 ha
Client: Communauté Urbaine de Strasbourg
Co-project manager: Egis France, Bécard & Palay
Photo credits: Atelier Villes & Paysages (n°02, 16), Balloïde-Photo (n°01, 08-11, 14, 15), L'Atelier Lumière (n°03)

01、02. 塑造一个让汽车进入其中的景观
03、04. 漫步空间的照明设计
01-02. The creation of a landscape where vehicles blend in
03-04. Lighting of the pedestrian paths

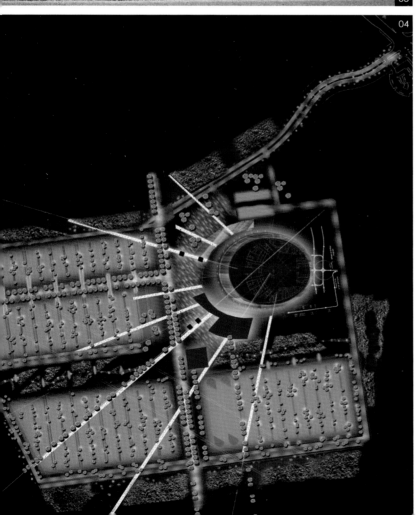

此项目涉及由建筑师马希米亚诺·福克萨斯所设计的斯特拉斯堡天顶音乐厅（具有1万个观众席）周边空间的整治，其中包括一个必须满足小汽车、摩托车、自行车和旅游客车等3500个车位的停车场。

这个主题乍看显得简单，实际上却颇为复杂：如何避免像大型超市的停车场那般让车位排排站？如何避免这些机动车造成过为显著的视觉效果？如何使其成为一个承载着音乐厅建筑的绿色珠宝盒？

整治方案围绕着三个主要的原则而进行构思：
- 打破一般停车场直角排列的逻辑
- 以"树林边界"所围成的小区来塑造不同的"停车蜂窝"，形成真正的空间区隔
- 以植物作为主题，发展出一系列的视觉轴线：乔木层、灌木层和小树林

The redevelopment of the surroundings of the Strasbourg Zénith, a 10,000-capacity concert hall designed by the architect Maximiliano Fuksas, involved finding parking spaces for 3,500 light vehicles, motorbikes, bicycles and cars.

This mission, which seemed simple at the outset, nevertheless remained complex: how to avoid lining the vehicles up like in an everyday hypermarket car park; how to limit their visual impact; how to plan this development with respect for an attractive natural site.

The development plan was based on three main principles:
- breaking with the orthogonal organisation of the car park,
- creating parking cavities, offering real spatial fragmentation, through "hedgerow cells",
- organising a development strategy around several readable axes based on a plant theme: the tree layer, the shrub layer and the hedgerow.

05-07. 依照植物层次结构而塑造的不同视觉轴线
08. 音乐厅前面一个可以提供节庆活动使用的林荫道广场
09. 厢房般的停车场区隔

05-07. Different ways to read the site: plant stratification
08. A forecourt and avenue on a festival scale
09. The parking lots

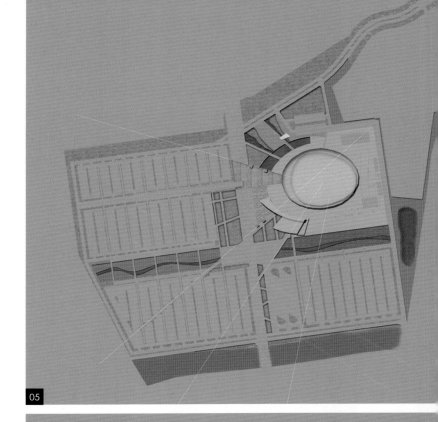

05

这是一个围绕着建筑而塑造的景观精品，一片"森林"，而天顶音乐厅就处于林中空地当中。建构停车场空间的植物如同一系列同心圆波浪，层层组成了以天顶音乐厅为中心的蜂房状网格。这些停车场以草沟和灌木丛作为分隔，而穿越其上的通道则将人们带往音乐厅的入口广场。由于雨水无法渗透停车场地面，因此整体的雨水收集透过了一系列的草沟来进行，将水引导到大型的防渗水池，这些水池同时也勾画出基地内的主要轴线。这个大型停车场由此变成了一个散步休闲的场所。

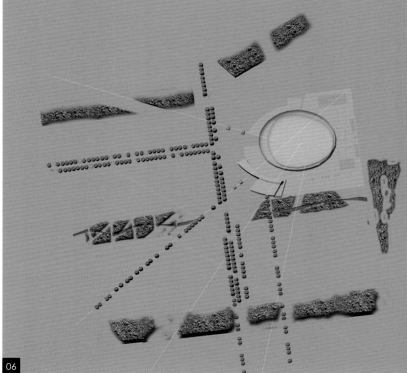

06

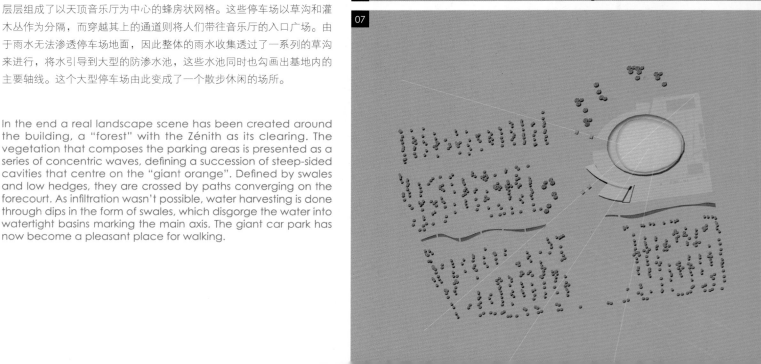

07

In the end a real landscape scene has been created around the building, a "forest" with the Zénith as its clearing. The vegetation that composes the parking areas is presented as a series of concentric waves, defining a succession of steep-sided cavities that centre on the "giant orange". Defined by swales and low hedges, they are crossed by paths converging on the forecourt. As infiltration wasn't possible, water harvesting is done through dips in the form of swales, which disgorge the water into watertight basins marking the main axis. The giant car park has now become a pleasant place for walking.

03 城市公共空间 Urban public spaces

10. 植物与硬质地面的结合方式采用了借自娱乐性公园的语汇
11. 排水系统的构思之一：经过绿化的集水盆地
12、13. 林荫道与停车空间剖面图
10. A plant and mineral vocabulary borrowed from pleasure parks
11. The drainage strategy: planted run-off collection basins
12-13. Cross-sections of the avenue and the parking spaces

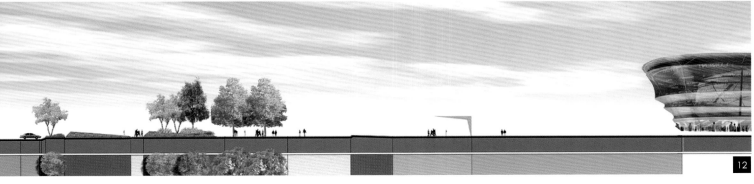

03 城市公共空间 Urban public spaces

14. 充满节庆氛围的夜间灯光
15. 照明设计强化了行人轴线，模糊了停车空间
16. 视觉轴线的集中点：天顶音乐厅

14. A festive approach to the lighting
15. The pedestrian paths are highlighted, while the parking lots are toned down
16. Target: the Zénith

03 城市公共空间 Urban public spaces

瓦兹河岸

Quays of the Oise

ATELIER RUELLE

地点：法国蓬图瓦兹
完工日期：2010
面积：14 000 m²
业主：塞尔奇-蓬图瓦兹城乡区域联合组织
照片版权：Gérard Dufresne

Location: Pontoise, France
Completion date: 2010
Area: 14 000 m²
Client: Communauté d'Agglomération de Cergy-Pontoise
Photo credits: Gérard Dufresne

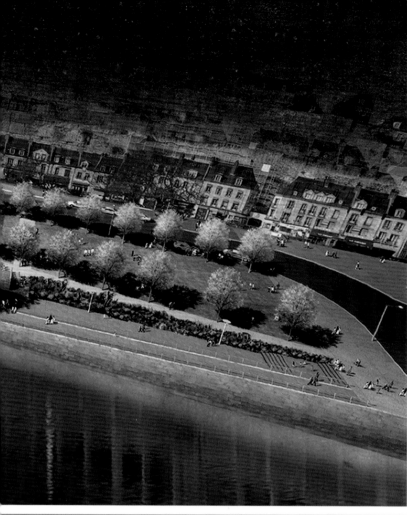

01. 方案透视图（2004年6月设计竞赛）
02. 位于老城墙和新游客中心（由Atelier Ruelle街巷工作室设计之建筑）脚下的一条散步道
03. 游客中心的屋顶成为观景台，可从高处堤岸直接抵达

01. Perspective drawing of the project (competition June 2004)
02. Riverside walk at the foot of the ramparts and new tourist office building by Atelier Ruelle
03. From the quay, the roof of the tourist office forms a lookout point

瓦兹河流经塞尔奇-蓬图瓦兹城乡区域，蓬图瓦兹城墙下的河岸改造属于逐步收复河岸的一系列计划之一。

基地处于老城中心的边缘地带，这片荒地已经成为停车场和杂草丛生的被遗忘的空间。设计师重新考虑了地面处理的方式，因而创造出一条伴随着河岸缓慢下降的回折坡道和一条散步道，一直延伸到连接了蓬图瓦兹和圣乌昂洛莫纳的一座桥边。在沿着河岸直线排列的梧桐树之间，设计师营造了一系列宽广而充满趣味的休闲空间。

方案设计在很大程度上考量了水流和垂直的城墙间的关联、墙体的节奏以及石材的呈现，因此旅游信息中心和水边栈台的建造自然而然地融入了基地之中，两者采用统一的铺地一直延伸到展览厅内部。信息中心的屋顶成为高处河岸的延伸和观景台，这个稍微倾斜的平台连接着通往城市中心的大街，把人们直接带到水流前。整个河岸改造都融合在这个被列为历史遗迹的保护基地中。

The development of the quays at the foot of Pontoise city ramparts forms part of a larger plan for the progressive reclamation of the banks of the Oise where it crosses the urban area of Cergy-Pontoise.

Bordering the old town centre, the site was abandoned, made up of brownfield, car parks or residual green spaces. The project proposes a new treatment of the ground surfaces in favour of gentle folds descending towards the water and a walk that extends to the bridge linking Pontoise to Saint-Ouen l'Aumône. It introduces spacious and fun leisure spaces between the rows of plane trees that line the water's course.

The construction of the tourist office building and a river transport stop slide into this composition, which is strongly influenced by the relationship between the water and the verticality of the ramparts, the lines of the walls and the presence of stone. The tourist office and the river stop share the same ground surface treatment, which runs all the way into the interior of the exhibition hall. The roof of the tourist office becomes an extension of the high quays and a viewpoint, while the water-level entrance is accessed via an esplanade with a gentle slope in prolongation of the street that descends from the town centre. The whole project fits into this remarkable site, which is classed as historic heritage.

04. 散步道旁的景观
05. 一个长条形花园
06. 游客中心里与外，两者采用相同的地面处理
07. 在高处堤岸和底处河岸之间的地面层次处理
08. 在旅客中心和河船驿站之间的堤岸散步道

04. On the riverside walk
05. A long garden
06. The tourist office "inside/outside" with a single floor surface
07. Folds in the ground between the high quay and the banks
08. The riverside walk with the tourist office and the riverboat stop

03 城市公共空间 Urban public spaces

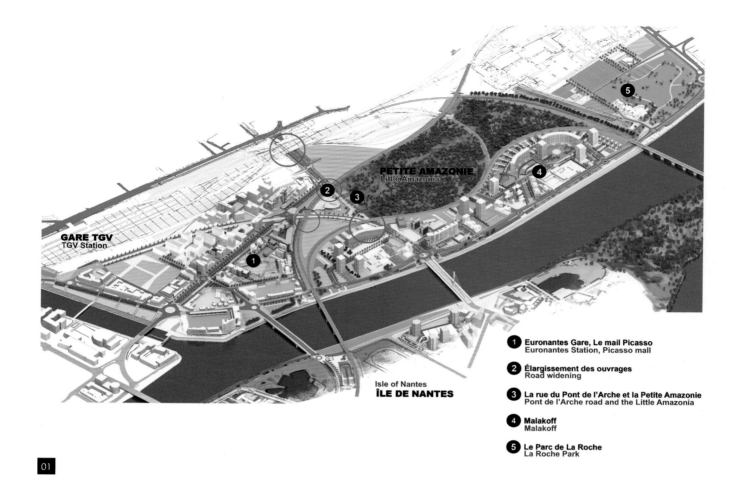

1. Euronantes Gare, Le mail Picasso
   Euronantes Station, Picasso mall
2. Élargissement des ouvrages
   Road widening
3. La rue du Pont de l'Arche et la Petite Amazonie
   Pont de l'Arche road and the Little Amazonia
4. Malakoff
   Malakoff
5. Le Parc de La Roche
   La Roche Park

01

欧洲南特火车站/马拉科夫
Euronantes Station/Malakoff

ATELIER RUELLE

地点：法国南特
完工日期：2001-2017
面积：164 ha
业主：南特大都会委员会和南特大都会开发机构
照片版权：Atelier Ruelle (n°02-05, 07-12), Gérard Dufresne (n°06, 13)

Location: Nantes, France
Completion date: 2001-2017
Area: 164 ha
Client: Nantes Métropole and Nantes Métropole Aménagement
Photo credits: Atelier Ruelle (n°02-05, 07-12), Gérard Dufresne (n°06, 13)

01. 方案全景
02. 马拉科夫街区向河流敞开
03. 新设学校对面的小型游戏广场

01. Overall view of the project
02. The Malakoff neighbourhood overlooks the river
03. A square facing the new school

欧洲南特车站/马拉科夫项目的核心目标就是在精神上和实质上连接位于市中心南部和南特车站南部的164公顷土地，它目前因为各种环境条件而产生隔离：地理的或者社会的分裂因素、铁路设施和飞地现状，这些都使人们产生强烈的疏离感。在东部，包含了1600户社会住宅的马拉科夫街区被孤立于铁道线和卢瓦尔河沿岸的快速路之间，需要被重新纳入城市之中。在西部，车站周围的荒地和"火车站配套"产业的遗留设施也存在着莫大的发展潜力。

本项目以扩展南特中心性为目标，把那些被忽略和弃置的街区纳入到城市中，提出一个新旧融合的发展计划，并且让城市与其显著的地理特色结合在一起：卢瓦尔河、小亚马逊湿地（Natura2000计划保护范围的湿地区）、拉罗什公园、圣·菲利柯斯运河……方案尽可能地把这些自然元素组合在一起，在连接西部和东部的同时也为所有公共空间创造出一个共通语汇。方案考虑了新的过街通道，组织城市中的主要干线，同时致力于提升邻近公共空间的质量。

The main objective of the Euronantes station/Malakoff project is to pull together – both physically and mentally – an area of 164 hectares south of Nantes railway station and the city centre. Several factors, among them the railway infrastructure and enclaved housing, had caused it to be disjointed, and it suffered from a strong sense of alienation. In the east, the Malakoff neighbourhood (1,600 units of public housing) trapped between the railway lines and the express road running alongside the Loire wants to become part of the city. In the west, the station surroundings, mainly made up of brownfield and station yards, offer a strong potential for development.

The project has to result in the extension of Nantes's city centre, incorporating neighbourhoods that have hitherto been left behind, to offer development integrating the old into the new, and open the city up onto a remarkable geography: the Loire, Little Amazonia (a wetland zone that is part of the Natura 2000 network), Laroche park, Saint Félix canal… It is permanently articulated around natural elements while generating a shared matrix of public spaces linking the east to the west. Its main lines are organised to form a link with the new pedestrian routes under the railway while paying constant attention to the quality of the public spaces in the immediate proximity.

04、05. 景观停车场
06. 毕卡索林荫道上的一景
07. 毕卡索林荫道是连结欧洲南特火车站和马尔科夫街区的东西向轴线
08. 毕卡索林荫道上的一景

04-05. Landscaped car park
06. Along Picasso mall
07. Picasso mall, an east-west link between Euronantes station and Malakoff
08. Along Picasso mall

03 城市公共空间 Urban public spaces

09、10. 从散步道可以看到受到保护的湿地区域
11. 位于东部、修复整治后的岩石公园
12. 罗莎·帕尔克街沿着公园伸展
13. 走向设有足球场的高地平台

09-10. The protected wetland zone is seen from the walk
11. In the east, the rehabilitated Roche park
12. Rosa Parks street running alongside the park
13. Going up to the terrace of the football pitch

03 城市公共空间 Urban public spaces

Three Rivers Mall
三河林荫道

ATELIER DE L'ÎLE – PAYSAGISTES / BERNARD CAVALIÉ

地点：法国斯丹
完工日期：2008
面积：18 000 m²
业主：法国93地区共同平原发展组织
照片版权：Atelier de l'île / Isabelle Otto

Location: Stains, France
Completion date: 2008
Area: 18 000 m²
Client: Plaine Commune 93
Photo credits: Atelier de l'île / Isabelle Otto

01. 整体平面配置图
02. 绿化斜沟的挖深设置是为了能够更长时间内地保留雨水
03. 在方案整治中，水存在的事实被刻意呈现出来

01. Master plan
02. Swales have been dug to retain the water for longer
03. The layout has been designed around the presence of water

三河林荫道是介于乔治·瓦尔邦公园、现有街区和新住宅区之间的公共空间。它在近乎700米长度上为周边居民提供临近的交流空间和通行空间：它是休闲游憩场所、街区之间的软性连接（可通往各种公共设施），也是通向省级公园的散步道。这片可吸收洪泛的区域使三河街区的雨水得以获得较好的控制和管理。

降雨在此方案中并不被视为是需要尽快排除的意外事件，反而被看作是基地内的活力元素：每个空间将随着降雨情况而以不同方式被雨水淹盖。草坪斜沟并不是受淹区域唯一的处理方式，设计师根据不同空间的环境背景和潜在的使用功能而采取不同的处理方式。林荫道上的各个空间，大草坪、湿地或可吸收洪泛的硬质小广场，都被设计成贴近自然的、拥有不同使用功能和气氛的空间。

Three Rivers mall is first and foremost a public space acting as an interface between Georges Valbon park, the existing neighbourhoods and the new residential fabric. Along almost 700 metres, it offers the inhabitants new outdoor spaces close to their homes and creates links: places for relaxation and recreation, soft links between neighbourhoods (access to facilities), a walk to the departmental park. A floodable space, it allows for the stormwater management of the Three Rivers neighbourhood.

Heavy rain is not seen as an accident to be evacuated as quickly as possible, but as an element to enliven the site: the spaces created flood according to the amount of rain, each in its own way. Grass-covered swales are not the only possible way of laying out these floodable spaces, and they have been treated differently according to their immediate context and the possible ways they will be used. Thus, along the soft link, wide meadows, a wetland zone or floodable hard-surfaced squares have been designed as spaces close to nature, with different ambiances and uses.

04. 可以吸收洪泛的游戏草坪是公园向城市延伸的部分
05. 步行小径确保了在公园中行进的连续性
06. 铺设草地的大型斜沟为人们提供了额外的使用空间

04. The floodable play lawn extends the park into the town
05. Pedestrian paths ensure continuity for walkers
06. Large grassy swales offer a continuity of use

为了避免雨水的淤积不利于林荫道的使用，方案所采用的技术措施需要适合空间环境和实用需求。那些人们经常使用的场所（草地、波浪广场）地势变化平缓，偶尔会遭到水淹但又可以很快地排干；反之，某些低陷空间则被深入挖凿且局部防止渗水，以便可以更长久地保留雨水。游戏草坪、植被斜沟、硬质广场以及湿地主要是提供居民消遣娱乐的实用场所。这些场所随着降雨情况而产生变化，使生活在高密度城市环境中的人们也能重新与自然接触，并且认识水在城市中的不同表现形式。

Not wishing to stock the stormwater to the detriment of the users of the mall, the landscape architects had to find the best way to adapt the technical solutions to the intended ambiances and uses. Thus, the spaces that are most heavily frequented (the lawn, the wave square), with soft forms, flood less often and dry out very quickly, while the low point of the site has been deeply dug out and partially waterproofed in order to retain the water for longer. Playing fields, planted water meadows, hard surface squares and wetland zones are above all spaces for pleasure and use. Evolving according to the amount of rainfall, they allow people to rediscover a contact with nature even in a dense urban environment, and diversify the presence of water in the city.

03 城市公共空间 Urban public spaces

07. 这个湿地区段经常遭到洪水的淹没
08. 禾本植物花园以其强烈的图案色彩点缀着散步道
09. 穿越性平台将人行步道连接了起来以保证步行的连续性
10. 位于住宅街区中心的砖铺地广场
11. 一个充满趣味而且可以从各个方向穿越的广场
12. 地面呈波浪状的硬质铺地小广场，雨后积水效果图

07. This sequence on the "wetland zone" is frequently flooded
08. The graphic forms of the grasses garden punctuated the walk
09. Raised walkways ensure the continuity of pedestrian links
10. The little brick square at the heart of the residential neighbourhoods
11. A fun square and a crossroads in every sense
12. Drawing of the wave-like form of the mineral plaza square

03 城市公共空间 Urban public spaces

阿里斯蒂德·白里安林荫大道
Aristide Briand Avenue

ATELIER DE L'ÎLE – PAYSAGISTES / BERNARD CAVALIÉ

地点：法国努瓦西-勒-格朗
完工日期：2006
面积：27 200 m²
业主：努瓦西-勒-格朗市政府
合作设计师：Atelier de l'île – Architectes / Brard & Le Bras-Quelen
图片版权：Atelier de l'île / Bernard Cavalié (n°01, 03-07, 09-11, 14-15), Diluvial / Hervé Devantoy (n°08, 12-13), Philippe Drancourt (n°02)

Location: Noisy-le-Grand, France
Completion date: 2006
Area: 27 200 m²
Client: Noisy-le-Grand City Council
Co-project manager: Atelier de l'île – Architectes / Brard & Le Bras-Quelen
Image credits: Atelier de l'île / Bernard Cavalié (n°01, 03-07, 09-11, 14-15), Diluvial / Hervé Devantoy (n°08, 12-13), Philippe Drancourt (n°02)

01. 位于市政府轴线上的下沉花园全景
02. 阿里斯蒂德·白里安林荫大道透视图
03. 方案与所在街区，整体平面配置图

01. Overall view of the sunken garden as an axis extending from the Town Hall
02. Perspective view of Aristide Briand Avenue
03. Master plan of the project and its localisation in the neighbourhood

在全面重振城市中心的整体计划框架下，法国努瓦西-勒-格朗市政府希望通过重新整合主要公共空间的使用功能而提升其价值（包括建设一个前庭广场、一座花园，扩大集市广场……），在四分之三的整治范围内建立人行系统，在市政厅和文化中心这两座处于城市心脏地带的重要机构之间建立一条结构性轴线，并强化建筑的展现。

阿里斯蒂德·白里安林荫大道侧面两排整齐的朴树强调了街道的轴线，形成具有强烈透视感的视觉效果。主要车行道位于大道的一侧，中间区域则处理成下沉花园，另一侧伴随着一条宽广的步行道。两座过桥让人们能够穿越花园，并且保证了林荫大道两侧的连通。

As part of an overall programme for revitalising its town centre, Noisy-le-Grand wished to improve its main public spaces through reorganising their uses (creating a forecourt and a garden, enlarging the market square...), creating a significant pedestrian zone on three-quarters of the surface area to be developed, improving the backdrop to the buildings and structuring the axis between two public establishments that symbolise the town centre: the Town Hall and the cultural centre.

The axial composition of the layout of Avenue Briand is supported laterally by two rows of hackberry trees that frame the view along a strong visual axis. The main road is moved to one side, while the central part becomes a sunken garden with a wide pedestrian walk running alongside. Two footbridges allow pedestrians to reach the garden and link the two sides of the avenue.

04. 方案轴线组织示意图
05. 方案材料使用示意图
06. 方案植被选择示意图
07. 方案水景设计示意图
08. 下沉花园上的栈桥
09. 下沉花园沿岸的人行道

04. Sketch of the layout approach for the axis of composition
05. Sketch of the materials for the project
06. Sketch of the planting for the project
07. Sketch of water in the project
08. Footbridge over the sunken garden
09. Pedestrian path along the sunken garden

"高处"段落通过下沉花园表现其特征，其中种植着多年生植物和禾本植物，点缀着小树丛。文化中心前一块大型毛糙石灰岩构成的喷泉成为景观亮点，喷泉水集合而成的水渠沿着花园的轴线依靠重力顺势流淌，展现出林荫道的坡地特征。"低处"段落则从有遮顶的集市大厅开始，花园在此中止，取而代之的是一大片红砖铺地的人行广场，这里经常被市集、旧货市场、节日集会所占据……水在坡道低处的市政厅立面之前又重新显露出来，以大型水池和水柱式喷水池的形式呈现。

The "high" sequence is marked by the presence of the sunken garden, planted with perennials and grasses and punctuated by attractive young trees. A stream tracing the axis of the garden begins in front of the cultural centre, emerging from a fountain in the form of a raw limestone block. Running as gravity takes it, it enhances the topographical perception of the sloping avenue. The "low" sequence begins in front of the covered market: the garden is interrupted, providing a wide, open pedestrian space paved in bricks, which is often used for funfairs, antiques markets, festivals... The water emerges again at the bottom of the slope in front of the Town Hall facade, in the form of a large ornamental lake and fountain.

03 城市公共空间 Urban public spaces

ESPLANADE DE LA FONTAINE-SOURCE   ESPLANADE MICHEL S

JARDIN CREUX

10

ESPLANADE DE LA FONTAINE-SOURCE   ESPLANADE MICHEL SIMO

JARDIN CREUX ET FIL D'EAU

11

10、11. 喷泉环境剖面图
12. 透视抽线被水渠强化出来
13. 文化中心前面的喷泉由大型毛糙石灰岩构成，四周环绕着水柱
14. 水渠的出发点
15. 花岗岩水渠细部

10-11. Sections of the fountain
12. The perspective axis defined by a stream
13. Fountain composed of a limestone block surrounded by water jets in front of the cultural centre
14. The beginning of the stream
15. Detail of the stream's granite bed

03 城市公共空间 Urban public spaces

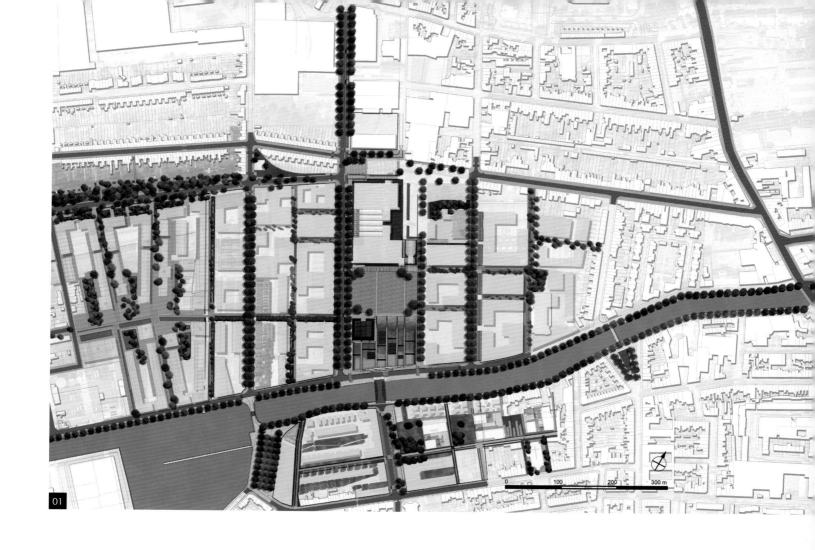

01

Banks of the High Deûle Eco-neighbourhood
上 德 勒 河 岸 生 态 街 区
ATELIER DE PAYSAGES BRUEL-DELMAR

地点：法国里尔
完工日期：2008-2015
面积：25 ha
业主：SORELI 混合经济开发公司，代表里尔市及里尔与洛姆城镇之城市联合组织
照片版权：Atelier de paysages Bruel-Delmar

Location: Lille, France
Completion date: 2008-2015
Area: 25 ha
Client: SEM SORELI for the communauté urbaine de Lille and Cities of Lille and Lomme
Photo credits: Atelier de paysages Bruel-Delmar

01. 整体平面配置图，一个转向水面的街区
02. 欧洲科技园区前面一条收集雨水的沟渠
03. 以植物来进行环境调节

01. Master plan, a neighbourhood based on water
02. One of the canals for rainwater harvesting in front of "Euratechnologies"
03. Phytoremediation

上德勒河岸协议开发区的改造立足于对基地环境品质的认识，以这些环境特质来建立方案的基础和丰富性。无论是在这个街区的历史中还是在当今的规划配置中，水体的存在都是毋庸置疑的，尽管人们逐渐失去了对它的认知。上德勒河岸方案就是循着这个有关水的记忆与痕迹而发展，以便延伸使用场所的识别性，并且在新的改造计划中更新水体的表达方式。内河航运业的遗迹强化了白树林街区的识别特征，其水运站前的广场经过整治而成为人们交流的场所。水生花园扮演着雨水储存和植物调节的角色，它的形态伴随着雨水的节奏而发生变化，成为开发区内水体设计的标志性场所。

被改造成欧洲科技园区的旧布朗·拉丰纱厂是这个街区的关键点。规划方案在这个建筑群四周设置了公共空间，南面有大草坪和水生花园，北面则有旧节日大厅和布列塔尼广场。这个供高科技企业使用的建筑和那些对大众开放的公共空间之间产生了某种交互作用和互补性。

上德勒河岸的发展为洛姆沼泽地区的公共空间提供了重新定位与整治的机会，在保留这个街区自身特征和魅力的同时，也为它提供新的交通方式。这条花园道路展现了德勒河谷的土地尺度，并且提出一种线性公园的形式，以连接整个街区的核心区，同时也尊重工人住房的形式。这个项目获得了2009年法国生态街区大奖的水主题杰出方案奖。

The Banks of the High Deûle ZAC (comprehensive development zone) takes a developmental approach of recognising the qualities of the site that forms the basis of the project and then enriching it. Water has always been present in this neighbourhood, both historically and in its present-day configuration, though it is less valued today. The project for the banks of the High Deûle takes the memory of the past as a starting point to provide continuity in the identity of inhabited places, and to give water a strong federating role again in the new development. The water station square offers a convivial outdoor space and a reminder of the inland water shipping that forged the identity of the Bois Blancs neighbourhood. The water garden, which plays a storage and phytoremediation role, evolves to the rhythm of the rains and has become the symbol of this water-related project.

The old textile mill Le Blan-Lafont, now "EuraTechnologies", forms the focal point of the neighbourhood. The project creates a link between this collection of buildings and the public spaces that border it, the large lawn and the water garden in the south, the old dance hall and Bretagne Square to the north. An interactive and complementary relationship is forged between the building, which houses hi-tech companies, and the public spaces open to all.

The development of the banks of the High Deûle brings a new use to the public spaces of Marais de Lomme and manages to improve the quality of service they offer without depriving the neighbourhood of its character and charm. This "garden-street" reflects the character of the Deûle valley as a whole, and with its linear form is a park that serves the heart of the neighbourhood while respecting its industrial past. In 2009 this project was given the Eco-neighbourhood label for its work with water.

04. 城市网络受到雨水收集网络的支配
05. 为西边街区突围
06. 水生花园提供了丰富且多样的生态环境
07. 在水生花园一个沙龙之前的萍蓬草和慈姑
08. 大草坪和水生花园成为街区之间新的连接空间

04. The urban canvas is dictated by the needs of rainwater harvesting
05. Opening up the western neighbourhoods
06. The water garden is ecologically rich and diversified
07. Nuphar pumila and Sagittaria in front of one of the rooms of the water garden
08. The main lawn and the water garden, new linking spaces between the neighbourhood

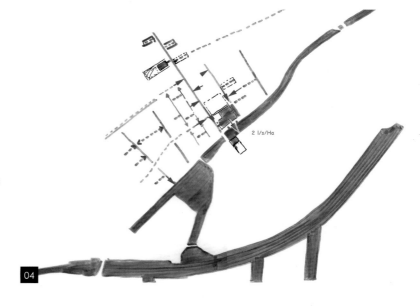

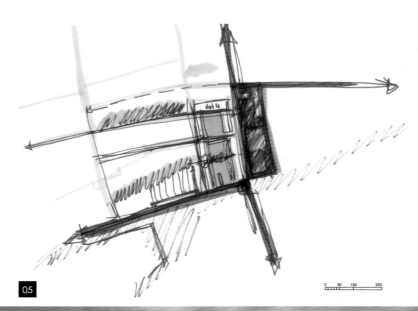

03 城市公共空间 Urban public spaces

09. 生锈的金属令人联想起基地往昔的工业历史
10. 沿着欧洲科技园区朝着水的方向走去
11. 布列塔尼庭院中由工业水泥板与熔化玄武岩组成的铺地
12. 一个向大众开放的空间，犹如一个新的分享场所
13. 与街区建立关系
14. 采用工业材质

09. Oxidised metal evokes the site's industrial past
10. Past "Euratechnologies" and down to the water
11. Industrial concrete slabs and fused cast basalt paving stones for Bretagne Square
12. Open to all, this is a new shared space
13. Creating links to the neighbourhoods
14. Industrial materials

03 城市公共空间 Urban public spaces

15. 新开辟的柳树大道为洛姆沼泽街区突破了重围
16. 一条新的"花园道路"
17. 让地面在提供使用性的同时也能够呼吸
18. 以多种方式来收集雨水
19. 柳树大道，一条融入历史印记中的花园道路

15. The new Saules Avenue opens up the Marais de Lomme neighbourhood
16. A new "garden-street"
17. Letting the ground breathe while allowing it to be used
18. Several methods of rainwater harvesting
19. Saules Avenue, a garden-street that follows historic lines

03 城市公共空间 Urban public spaces

Charles de Gaulle & Saint Vaast Squares
戴高乐广场与圣瓦斯特广场
ATELIER DE PAYSAGES BRUEL-DELMAR

地点：法国阿尔芒蒂耶尔
完工日期：2008-2010
面积：30 000 m²
业主：里尔大都会城市联合组织、阿尔芒蒂耶尔市政府
照片版权：Atelier de paysages Bruel-Delmar (n°06-13), Max Le Rouge (n°01, 03, 05)

Location: Armentières, France
Completion date: 2008-2010
Area: 30 000 m²
Client: Lille Métropole Communauté Urbaine, Armentières City Council
Photo credits: Atelier de paysages Bruel-Delmar (n°06-13), Max Le Rouge (n°01, 03, 05)

1	Course of the river Lys
2	'Vivat' Theater
3	'Vivat' Garden
4	St -Vaast Church
5	Church forecourt
6	Town hall
7	Charles de Gaulle Square
8	St-Vaast Square
9	Dunkerque Street

01. 市政府前重新整治过的广场
02. 整体平面配置图，这是一个让城市历史重新显迹、新旧重叠的方案
03. 石块铺地的线条为广场空间建立起新的秩序，而水柱喷泉则为城市居民提供了生活趣味

01. The reunified Grand'place in front of the Town Hall
02. Master plan of a palimpsest project for rewriting urban history
03. Lines in the stone order the space. The fountain contributes to the life of the city centre

在阿尔芒蒂耶尔，急剧唐突的事件不断层叠和并置所造就出的动荡历史为城市留下空旷的空间，城市中心重新定位和整治的关键就体现在对这个现有实况的处理之上。为了同时塑造方案和历史，设计师借助"隐迹文本"的概念，以当代的手法将基地重新创造成为一个具有意义的场所。方案因此围绕着两个基本点展开：以基地特性作为基础，以使用功能作为调控元素。

每个空间都围绕着一个市中心的主要建筑而发展：市政厅、圣瓦斯特教堂以及如今变成维瓦特剧院的旧市场。沿着敦刻尔克街伸展的戴高乐广场是将昔日的大广场、市政府前庭和逝者纪念碑前空间重新整合所形成的广场，通过一些组织线条和来自于弗朗德勒的石灰岩铺地而达成整体空间的统一。

与其相对的圣瓦斯特广场是戴高乐广场和利斯河旧河床之间的过渡空间。景观师利用这个情势而将它设计成介于广场和花园之间的综合体，在城市中心为人们提供一片休闲空间，让人们联想起往昔的"手球场"或者"玩滚球的草坪"。在维瓦特剧院和教堂周围一些旧空间的"轮廓"被一一恢复，塑造成广场和周围围绕着矮墙的平台，透过三道宽敞的坡道将人们引导到将被改造成绿色散步道的利斯河旧河床。

In Armentières, a lively geological history of stratifications and brutal juxtapositions has led to a spatial vacuum that forms the main challenge for the redevelopment of the town centre. It was decided to reinvent the place in a contemporary way, incorporating this history into the project in the manner of a palimpsest. Thus the project is built around two fundamental points, the site as foundation, and the uses as governing ideas.

Each space is articulated around one of the major buildings of the town centre: the town hall, the Saint Vaast church and the old market hall, which is today the Vivat theatre. Général de Gaulle Square, opening out from Rue de Dunkerque, has been reformed by bringing together the old Grand' Place, the area in front of the town hall and the war memorial. It finds its unity in its lines and its surface of Flanders limestone.

In counterpoint, Saint Vaast Square is at the junction between Général de Gaulle Square and the old Lys riverbed. Making the most of this situation, it aims to be a hybrid square and garden, recalling the old *ballon au poing* grounds or bowling greens, to offer a space for relaxation in the heart of the town. The "contours" are restored and create the forecourts and the low-walled terraces around the Vivat and the church, orienting the space via three wide ramps pointing towards the old Lys riverbed, which will become a green walk.

04. 四个与城市历史和地理息息相关的发展基础：
大广场和花园广场；从城堡到街道与街坊；利斯河旧河床；介于城市逻辑与自主性之间的市政府建筑
05. 城市中心
06. 在棠棣树下找到属于行人的尺度
07. 露天市集使用的水槽
08. 水是方案构思的引导线

04. Four foundations based on history and geography: the Grand'place and the garden square; from the fortifications to the built blocks and the streets; the old Lys riverbed; the Town Hall mid-way between urban integration and autonomy
05. The town centre
06. Reverting to a pedestrian scale under the Juneberry trees
07. The market fountain
08. Water as a guiding line

03 城市公共空间 Urban public spaces

09. 花园广场
10. 圣瓦斯特广场，一个绿化的公共开放空间
11. 一些旧空间的"轮廓"被一一恢复，通过矮墙和宽阔阶梯的形式呈现
12. 红砖矮墙体现出空间轮廓
13. 红砖墙上"落水管弯头出口"的象征性处理体现着"水流路径"

09. The garden square below the church
10. Saint Vaast Square, a planted public space
11. The "contours" are rebuilt in the form of low walls and steps
12. Contour-forming brick wall
13. A drainage opening in the brick wall reminds us of the "water course"

03 城市公共空间 Urban public spaces

波提耶尔-申内生态街区

Bottière Chênaie Eco-neighbourhood

ATELIER DE PAYSAGES BRUEL-DELMAR

地点：法国南特
完工日期：2008-2015
面积：30 ha
业主：南特大都会开发机构
照片版权：Atelier de paysages Bruel-Delmar

Location: Nantes, France
Completion date: 2008-2015
Area: 30 ha
Client: Nantes Métropole Aménagement
Photo credits: Atelier de paysages Bruel-Delmar

01. 整体平面配置图，一个与地块历史关系紧密的生态街区
02. 勾哈溪被重新展现于大自然之中
03. 一座风车抽取水来灌溉新设的菜园

01. Master plan, an eco-neighbourhood rooted in the land parcels of its agricultural history
02. The Gohards stream uncovered
03. A windmill draws water to irrigate the new vegetable gardens

波提耶尔-申内街区的整治被纳入到基地的历史和地理环境中。其历史的突出之处在于人们对这片土地的使用，过往的地质环境在基地内形成的受压蓄水层易于打井，因此居民们利用了这个天然条件在城门附近开垦菜地。

方案企图将这些历史元素表达出来，以便使项目能够在基地扎根并成为一个独特的场所。公园和公共空间被融入过去农田的地块结构中。方案的具体形态建立在基地中现存墙体的线条之上，犹如一个沉淀出来的结果。呈东南、西北走向的公园在新林荫大道的衬托下强化了老杜龙街区和波提耶尔街区之间的联系。公园也起到平衡新密度的作用，其绿化以本土植物作为主导，呈现出一种被重新找回的自然。

The project for the development of the Bottière Chênaie neighbourhood is deeply rooted in the history and geography of the site. With a geological past that has left confined aquifers enabling the sinking of wells, human beings have long taken advantage of this situation to grow fruit and vegetables in parcels on the outskirts of the city.

The project involved researching these aspects in order to anchor it in the site, and to situate it as a unique place. The park and the public spaces follow the network of land parcels belonging to its agricultural past. Like a work of sedimentation, the new design follows the lines of the existing walls. The park, oriented south-east north-west, asserts a new link between the old neighbourhoods of Doulon and Bottière while running alongside the new mall. An urban park balancing the new densities, its primarily indigenous vegetation will give the impression of nature rediscovered.

04. 街区和公园之间的关系围绕着溪流而组织
05. 勾哈溪的重现对街区居民的生活产生重要的影响
06. 勾哈溪公园的规划借助于这条重新找回的溪流
07. 一条用来收集雨水的深沟
08. 沿着溪流散步
09. 公园里的风车
10. 坐在草地上休息成为稀松平常的事

04. A link between the neighbourhoods, the park is organised around the stream
05. The uncovering of the Gohards stream influences the life of the neighbourhood
06. The new Gohards park around its rediscovered stream
07. A large swale harvests stormwater
08. A walk alongside the stream
09. The park's windmills
10. Sitting in the grass is a daily activity

方案刻意重现勾哈溪的做法是为了回应场所的记忆，特别是对地理形态的记忆。此溪流成为整个公园的重心，并且接收新街区的草沟和水渠所带来的雨水。在这块土地上展现了诸多人类智慧的痕迹，其中一些与水井连接的灌溉用蓄水池构成了、也延伸了此地块的识别特征。除了把这些蓄水池保留下来并进行必要的整修以便使用，风能也被应用在灌溉系统中，为未来的共享花园（家庭式花园）和公园服务。

Uncovering the Gohards stream is a way to express the memory of the place and in particular its geography. It becomes the heart of the park and the place for harvesting the stormwater of the new neighbourhood, channelled by swales and canals. Among the numerous traces of Man's intelligent use of the site are several reservoirs for stocking water for irrigation. Linked to wells, they remain and form part of the identity of the place. Aside from their conservation and in some cases their restoration, wind energy has been harnessed to put them to use once again for watering the future community gardens and the park.

03 城市公共空间 Urban public spaces

11. 在此街区，走在马路正中间成为一件良好的习惯
12. 在生态街区里，道路旁的空间被用来收集雨水
13. 新住宅入口的过桥
14. 绿化的街道为空间带来清凉
15. 住宅区道路剖面图，每条道路必须有50%的绿化面积

11. Here, walking in the middle of the road is safe
12. In the eco-neighbourhood, stormwater is harvested along the roadside
13. Approaches to the new residential blocks
14. The planted streets make the place seem fresher
15. Cross-section of a residential street, each of which is planted over 50% of its surface area

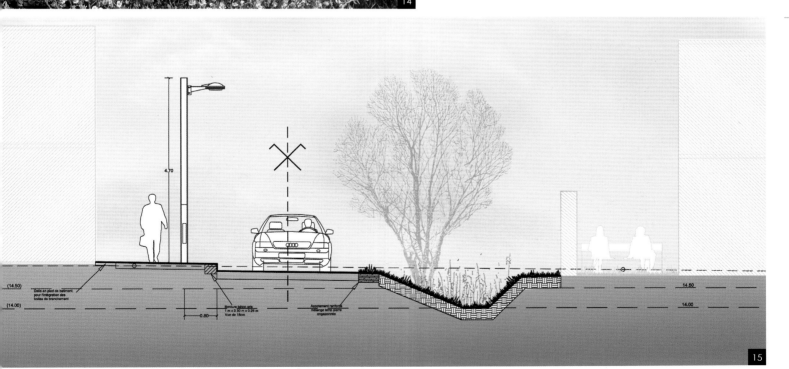

03 城市公共空间 Urban public spaces

16. 这条水渠收集和储存雨水
17. 这条流经学校前面的水渠，为各街区之间建立了明显的关系
18. 坐在水边是日常生活中经常出现的事情
19. 在库斯托广场上，一棵老桉树被保留了下来
20. 洗脚池形式的空间同样也是用来管理雨水的方法之一

16. The canal harvests and stocks stormwater
17. In front of the school, the canal marks the strong link between the neighbourhoods
18. An everyday scene, sitting by the water
19. On Cousteau Square, an old eucalyptus from a garden has been preserved
20. The paddling pool is also a tool for stormwater management

03 城市公共空间 Urban public spaces

21、22. 榭诗里街区，可渗透的地面和可储水的空间
23. 在一个被保留的老农场和新住宅之间
24. 要进到室内，首先得穿越绿化斜沟，并且经过被绑缚的幼株梨树
25. 通往公园的一条步行小径和其旁的绿化斜沟

21-22. In the Sècherie neighbourhood, permeable surfaces and storage areas
23. Between an old farm that's been preserved and the new housing
24. To enter, cross the planted swale and go past the trellised pear trees
25. A pedestrian path and its swale lead to the park

03 城市公共空间 Urban public spaces

26. 寇林纳街区的分享式花园面对着住宅分布
27. 寇林纳街区的分享式花园属于勾哈溪公园的一部分
28. 分享与交流的场所
29. 孔潘尼庸盆地剖面图：一个下雨时候可以用来汇集雨水的干燥结构体
30. 利用地面空间来管理雨水有利于塑造多样化的生态场所
31. 以蓝页岩石笼作为墙体
32. 以橡木作成的阶梯

26. The Collines allotments facing the new housing
27. The Collines allotments belong to the Gohards park
28. A place for sharing and conviviality
29. Cross-section of the Compagnons basin, a normally dry structure designed for harvesting stormwater
30. The surface management of water encourages the creation of places for biodiversity
31. Blue shale gabions
32. Oak steps

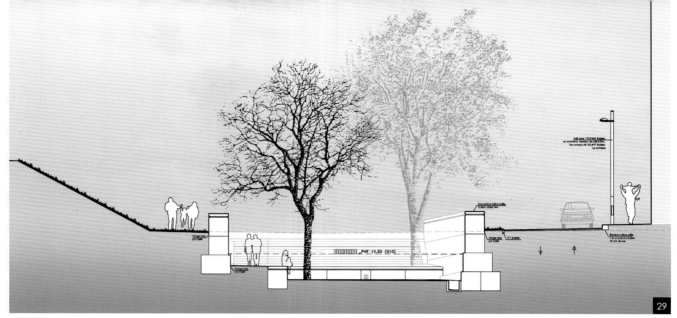

03 城市公共空间 Urban public spaces

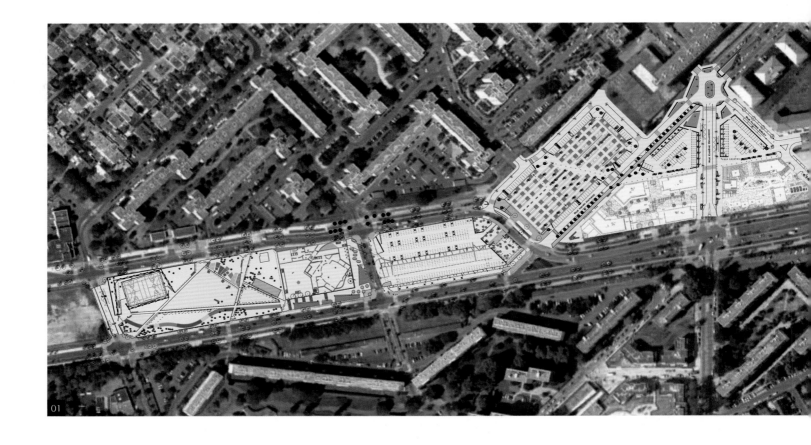

奥奈协议开发区 *Aulnes ZAC*

COULON LEBLANC & ASSOCIÉS

地点：法国奥奈丛林
完工日期：2012 – 第一阶段；2013-2014 – 第二阶段
面积：2.5 ha – 第一阶段；3 ha – 第二阶段
业主：Sequano Aménagement混合经济开发公司
照片版权：Jacques Coulon

Location: Aulnay-sous-Bois, France
Completion date: 2012 – 1st phase; 2013-2014 – 2nd phase
Area: 2.5 ha – 1st phase; 3 ha – 2nd phase
Client: Sequano Aménagement
Photo credits: Jacques Coulon

01. 整体平面配置图：公共花园、清真寺、停车场、露天市场、停车场
02. 公园西南边入口
03. 公园的中央道路与植被，玫瑰和锦带
04. 游戏围栏和硬木座椅

01. Public garden, mosque, car park, market square, car park
02. South-west entrance to the park
03. The central walk and beds of roses and weigela
04. Games cage and hardwood benches

奥奈丛林这个城市原先被高速公路及其两旁的保护空间切成两个部分，历经几十年的重新整治才慢慢将市区再度结合起来。如今城市由东到西呈现出一个具有连续性的空间，其中设置了一个公园、一个运动空间、一个停车场、一个露天传统市场用地、若干住宅与产业建筑，甚至还有一个清真寺。艾梅·塞泽尔公园和停车场被设置在原先过渡性车道所遗留下来的植物废墟空间里，其中某些俊秀的树木被保留了下来，公园里的活动空间包括一些花园、几个法式滚球场地和一个综合球场，完全是计划中的产物，然而日常使用空间却不然。此公园呈长条带状，人们却在南北向斜线穿越。在这个项目进行构思的时候，这些斜向穿越的路径特别引起了设计师的注意。

空间围塑与界定的方式变化多样，当涉及视觉景观与行人空间时则较为开放，而涉及各种交通穿越时则具有选择管控性质。这些场地界线包括线状构筑的低矮石笼、排列疏松的金属栏杆、球场旁高而密集的围栏、由花岗岩叠成的长条厚实座椅和边界。一些非常坚实的金属围栏依据选择而允许不同交通模式的穿越：脚踏车、步行、婴儿推车与残障人士、维修公园的工程用车等，而避免机动车的经过。这些围栏不仅是技术性的工具，同时也是雕塑性的装饰。夜晚来临时，多彩的灯光在这些管控点的门栏上闪烁着，而公园其他部分的照明则以相当简约的方式处理。

Aulnay-sous-Bois used to be crossed by a highway that cut it in two. After it was closed, several decades of reconquest were needed for the town to sew itself back together. Today, it has an urban continuity from west to east, a park, sports grounds, a car park, a large market square, housing, businesses and, finally, a mosque. Aimé Césaire park and the car park are now installed in the strip of wasteland that contained the lanes of the highway. The most attractive trees were carefully preserved. The facilities of the park – a few gardens, boules pitches, a "cage" for ball games – were eagerly anticipated, but the residents had to be encouraged to use it as an everyday resource for getting from A to B. Although it is a long space, it tends to be used crossways, from north to south. Particular attention was paid to these cross paths at the design stage of the project.

The system of enclosures is both very open with regard to views and people, and very selective regarding modes of crossing the park: low gabion walls, very open metal fences, high, dense fences for the ball games, rows of thick granite benches that serve as borders as well as seats. Thick, indestructible metal gates make it possible to select bicycles, pedestrians, pushchairs and wheelchairs, as well as maintenance vehicles, while avoiding motorised two-wheelers. These filters are not only technical, but sculptural too. At night, these gates are lit up with coloured lights, while the rest of the lighting is in contrast very sober.

05. 南北穿越公园的对角线
06. 只允许步行进入的入口围栏
07. 停车场的人行步道，沿途设置了8'18''照明与造型设计公司的灯具
08. 公园一景，远处背景为游戏围场

05. The main north-south diagonal
06. Pedestrian gate
07. Pedestrian path through the car park, showing the work of the lighting artist 8'18''
08. The park with the games cage in the background

城市小品随着长长的小径和预期用途的密度而变化，以坚木制作、形体相同的建构物点缀着。停车场能够依据新露天市场的需要而提供汽车和卡车的混合使用，其限制高度的栅栏在技术性功能之外，也同样扮演着空间雕塑的角色，使人们从远处便能辨识它们。高大的路灯和地面的标志图案也都经过精心设计，使得停车场不仅仅是汽车使用的技术性空间和穿越空间，也成为建构生活品质和新街区意象的重要元素之一。

The furniture has been designed on the scale of the long walks and the expected density of use, using hardwoods and vandalism-proof construction. The car park is designed for mixed use by both cars and lorries, linked to the new market. Beyond their technical rôle, the lintels and height-limit bars also play a role as sculptures, recognisable from afar. The large candelabra-style lamps and graphic markings on the ground also play a part: the car park is not only a technical place linked to vehicles, but, crossed by a wide walk, it contributes to the quality of the lived environment and the image of the new neighbourhood.

03 城市公共空间 Urban public spaces

伊莲D与查尔斯A·萨蒙斯公园
Elaine D. & Charles A. Sammons Park

MICHEL DESVIGNE PAYSAGISTE

地点：美国达拉斯
完工日期：2009
面积：2.5 ha
业主：AT&T表演艺术中心
合作设计师：Foster + Partners, architectes ; REX/OMA, architectes ; JJR, LLC, Deb Mitchell
照片版权：Michel Desvigne Paysagiste (n°11, 12), Chris Heinbaugh (n°08), Carter Rose (n°03, 07, 09, 10), Nigel Young / Foster + Partners (n°01)

Location: Dallas, United States of America
Completion date: 2009
Area: 2.5 ha
Client: AT&T Performing Arts Center
Co-project manager: Foster + Partners, architects; REX/OMA, architects; JJR, LLC, Deb Mitchell
Photo credits: Michel Desvigne Paysagiste (n°11, 12), Chris Heinbaugh (n°08), Carter Rose (n°03, 07, 09, 10), Nigel Young - Foster + Partners (n°01)

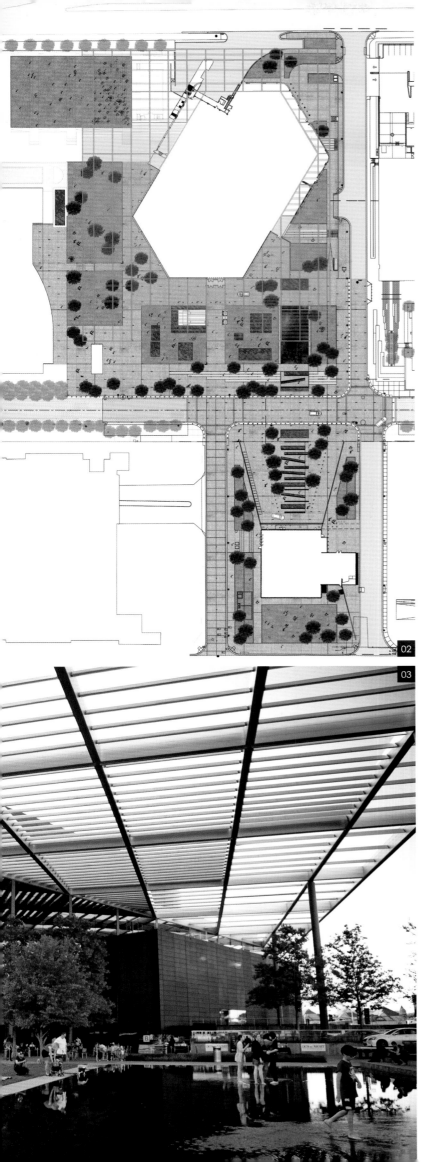

01. 萨蒙斯公园位于达拉斯市艺术区的街区中心
02. 整体平面配置图
03. 此公园是城市中心少见的能够聚集人群的空间

01. Sammons Park, in the heart of Dallas's Art District
02. Sammons Park, master plan
03. One of the rare gathering places in the city centre

美国达拉斯的艺术区是一个持续变化的街区，聚集了博物馆和新建筑：刚刚竣工的伦佐·皮阿诺设计的博物馆、诺曼·福斯特的歌剧院和雷姆·库拉斯的剧场。后两个建筑位于一片空地的两侧，业主以欧洲历史广场为蓝本作为参照对象，希望把它建设成为一个"中央广场"。

在这个位置稍微偏街区边缘、零星点缀着独栋建筑的大片沥青停车场地块上，以欧洲历史广场作为设计参考并不具有真正的意义。于是，设计师反其道而行，提出一种美国式规划网格的转换，使它成为一个抽象的基座，容纳着目前的和即将到来的各种不同元素：各种长方形花园、镜面水池、坡道和平台。它们以可移动元素的形式出现，随着财务状况和文化活动的需要而布置。

这个硬质铺地基座的构想首先以模型拼图的形式呈现出来，让参与设计者可以对各个元素的位置、比例和用途进行讨论。这个永远都可以进行重新组合的设置使广场能够随时抛开先前的组合形式，随着艺术区每个特殊情况的需要而进行调整。

The Arts District, a neighbourhood in constant transformation, groups together the Dallas museums and new buildings: a museum by Renzo Piano, an opera house by Norman Foster and a theatre by Rem Koolhaas. These two latter buildings are situated on different parts of a piece of derelict land, which the client wished to make into the "central square", based on the model of historic European squares.

In this slightly marginal place, defined by the asphalt surfaces of large car parks punctuated by isolated buildings, this reference didn't really have any meaning. Going against this idea, the proposition moved towards a transposition of the American urban grid, which becomes a kind of abstract plinth. This plinth hosts various objects, both present and to come: rectangular gardens, ornamental lakes, ramps, mobile terraces, all making themselves available to what financial and cultural opportunities come along.

A puzzle-model was built with this great asphalt plinth as its base, encouraging debate about the localisation, proportion and use of the different elements. Renouncing all forms of pre-established composition, this game of permanent recomposition allows one to accommodate whatever is happening in the District at a given time.

04-06. 研究模型：基本组成元素的排列组合
07. 植物性组成元素和大型硬质铺地广场的结合
08. 植物性组成元素和浅水池的结合

04-06. Study models: a play of juxtapositions of elementary components
07. Plant components and the large, neutral hard surface plaza
08. Plant components and the water feature

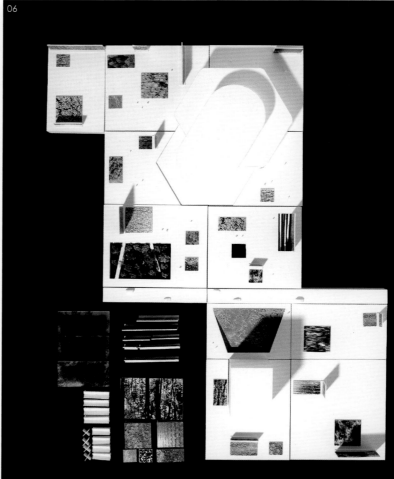

03 城市公共空间 Urban public spaces

09. 此公园容纳了温斯皮尔歌剧院的露天咖啡座
10. 介于广场、建筑前庭和公园之间的空间类型
11、12. 从温斯皮尔歌剧院大厅看到的户外景观

09. The park incorporates the terraces of the Winspear Opera
10. A typology that falls between square, forecourt and park
11-12. View from the lobby of the Winspear Opera

03 城市公共空间 Urban public spaces

奥斯特里茨广场
Austerlitz Square
DIGITALE PAYSAGE

地点：法国斯特拉斯堡
完工日期：2012
面积：9 900 m²
业主：斯特拉斯堡城市联合组织、斯特拉斯堡市政府
合作设计师：Architectures Amiot Lombard
照片版权：DIGITALEpaysage (n°02-05, 07-09, 11), Claudine Marchal (n°01), Bruno Steiner (n°10)

Location: Strasbourg, France
Completion date: 2012
Area: 9 900 m²
Client: Communauté Urbaine de Strasbourg and Strasbourg City Council
Co-project manager: Architectures Amiot Lombard
Photo credits: DIGITALEpaysage (n°02-05, 07-09, 11), Claudine Marchal (n°01), Bruno Steiner (n°10)

01. 一个"平面球形图",由众多小世界积聚成的一片星座
02. 从地面升高的带状元素让人联想起中世纪的老城门

01. A "planetary" square, a constellation of small worlds
02. The raised ground evokes the old Medieval gate

奥斯特里兹广场从前被称作屠夫门,它是老城和即将成为城市轴线的"赫利兹卡尔"之间的联结点,广场上总是人来人往,熙熙攘攘。这个项目的关键是要建设一个真正的生活空间,成为其所连接的四个街区的居民日常使用的场所。它同时也具有作为交通过渡空间的功能,因此在方案设计上不仅要考虑城市尺度,同时也兼顾街区和个体的尺度,并使"穿越"和"移动"成为花园构思的重点。

移动流线的印记被纳入到广场的空间组织当中;它如同一篇文字被刻印到相应的地面上。两种类型的空间和两种形式的使用方式在广场上并存:城市基座用来组织各种流线的移动和交错,而花园则是较为自由、较少束缚的空间,让人可以从城市中抽离片刻,在其中休憩与游戏。硬质铺地和城市小品的几何形状与圆形花园所象征的生命体有机形态两者在此和谐地交织在一起。

Austerlitz square, formerly the Butchers' Gate, is a busy junction between the old city and the future urban axis of "Heyritz Kehl". The challenge of this project was to create a truly living space that could be appropriated by the inhabitants of the four neighbourhoods that revolve around it, without denying its transit purpose. It meant working on the scale of the town but also on the scale of the neighbourhood and of the individual, and to make landscape and movement the conceptual raw materials of the project.

Movement is imprinted on the composition; like a text, it prints the scheme of our movements on the plinth that receives it. Two kinds of spaces and two kinds of uses cohabit here: the urban plinth, which is that of movement, of crossings and flux, and the garden, freer and less constrained, which allows people to escape from the city for a moment; a garden as refuge, a garden for play. The geometry of the mineral – of architecture – and the organic forms of the living symbolised by the circle-gardens cross paths here harmoniously.

03. "庇护"花园，开花的草坪上
04. 由附近居民管理的参与性花园以及其堆肥设置
05. 在这个由众多小公园组成的群岛中，可以发现各种不同的前进小道
06. 整体平面配置图：一个符合城市尺度的广场，也是提供给街区居民使用的花园
07. 混合栽种着不同层次的植物：乔木、灌木和草本植物

03. "Refuge" gardens, encircled and placed on carpets of flowering lawns
04. A community garden compost bin, managed by the local residents
05. Different channels for navigating through the garden archipelago
06. A square on the scale of the city, and gardens for the neighbourhoods
07. A green surface that mixes trees, shrubs and perennials

这个方案致力于让城市历史重新呈现出来，并且再度赋予旧屠夫门作为城市入口的地位。因此塑造了一条具有活力、能与城市产生和互动的边界，它同时也具有雕塑和建筑的双重特征，能够为人们提供小憩的场所：地面的天然石材透过一条记忆带而与老城衔接在一起。花园被安置在"城墙外"，在新设之城门所划定的象征性界线之外，此界线令人回想起斯特拉斯堡的老城墙。花园植物所带来的生命力从圆形几何图案里张显出来，并与城市性（即广场的硬质铺地）形成对比。

The aim was to reveal the history of the city and to give the old Butchers' Gate back its threshold status by creating an active and participative limit. It had to be sculptural and architectural, capable of concentrating an urban pause in its solidity: the memory stick from which the natural stone rejoins the old city. The gardens are established "extra muros", beyond the symbolic limit drawn by the new gate – evoking the old Strasbourg city walls. It is from these circles that the living expresses itself in contrast with urbanity, the hardness of the square.

03 城市公共空间 Urban public spaces

08. 在人潮频繁的中央轴线旁，设置了一个较为宁静的空间
09. 一个球状起伏的地面颇受青少年的青睐
10. 三棵被保留的大槐树为广场带来绿荫
11. 一个"线形沙龙"象征着历史城市的老城墙界线

08. A calmer space on the edge of the busier central axis
09. A surface that seems to bubble up welcomes youthful exuberance
10. Three superb sophoras that have been preserved lend their shade to the square
11. A "linear room" embodies the threshold of the historic city

03 城市公共空间 Urban public spaces

01

施托克教士广场
Abbé Stock Square
HYL

地点：法国沙特尔
完工日期：2010
面积：5 000 m²
业主：沙特尔市政府
照片版权：Arnaud Yver

Location: Chartres, France
Completion date: 2010
Area: 5 000 m²
Client: Chartres City Council
Photo credits: Arnaud Yver

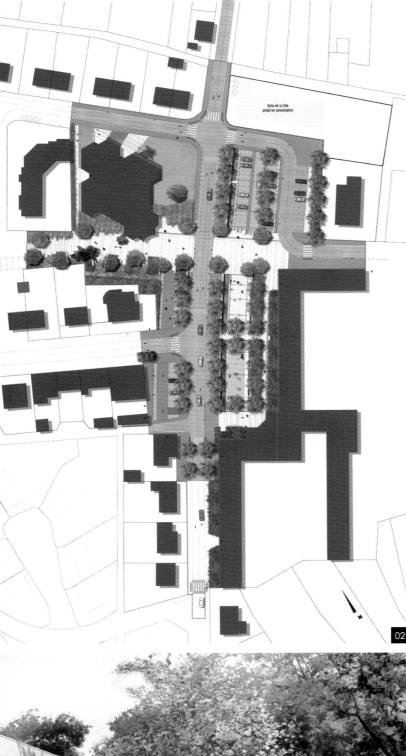

01. 停车场、教堂前庭和广场
02. 整体平面配置图
03. 教堂前庭透视图
04. 游戏场透视图

01. Parking area, forecourt and square
02. Master plan
03. Perspective drawing of the church forecourt
04. Perspective drawing of the playground

此广场的改造为这片战后诞生的别墅区带来了活力和舒适的环境。四条椴树林荫道的设置一下子就章显出这个场所的重要性，其广大的绿阴范围也为空间带来整体性。此空间通过花园广场的形式容纳着一系列具有不同功能的次级空间：学校附近的游戏场、隶属于教堂的安静花园、一些用于停车的小广场。

教堂和学校各自的前庭把广场与周围的城市肌理联系在一起。两个前庭汇集于一条贯穿广场的宽阔散步道，散步道的界限则消失在私人花园的边缘。一条长长的排水沟加上小灌木的厚度把游戏场和车行道间隔开来，游戏场虽然受到内在化的保护，却也同时向着大人的世界敞开。

交通轴线以柔和的方式穿过这个花园广场，以避免扰乱了其秩序。方案特别着重道路的材质与配备设施的设计，以便衬托出一旁存在的古迹遗产，并表达出这个人车共用空间的城市性。停车空间以砂岩铺地，标示边界的大块石灰岩如同乐曲的旋律一般变换着体量，它们时而犹如巨大的石凳，时而像一道道矮墙或者"庞贝城式的"防止停车的界石。

The laying out of the square consolidates and re-energises a small central area created after the War for the nascent residential neighbourhood. The addition of a four-fold mall of lime trees immediately gives the place a sense of importance. Its large leafy canopy unifies the space, sheltering a collection of sub-spaces within a garden square with various uses: a playground near the school, a quiet garden for the church, parking areas.

The forecourts of the church and the school link the square to the surrounding urban fabric. They are brought together by a wide walk crossing the square, whose outer edges merge into the private gardens. Protected from the traffic by a long ditch bordered by shrubs, the playground square looks inwards while still opening up to the world of grown-ups.

The traffic axis slips gently across the garden square without upsetting its order. Great importance has been given to the materials and treatment of the road, to enhance the existing heritage and express the urban nature of the community space. The parking areas are paved in sandstone. Blocks of solid limestone form a leitmotif that marks the edges of the spaces in the form of wide benches, a long, low wall or "Pompeian" no-parking bollards.

05. 教堂前庭延伸到道路的另一侧
06. 从停车场看向广场
07. 路边的石头犹如音乐节奏
08. 教堂花园
09. 教堂前庭的边缘处理

05. Road crossing the forecourt
06. The square from the parking area
07. The music of the stones
08. The church garden
09. The borders of the forecourt

03 城市公共空间 Urban public spaces

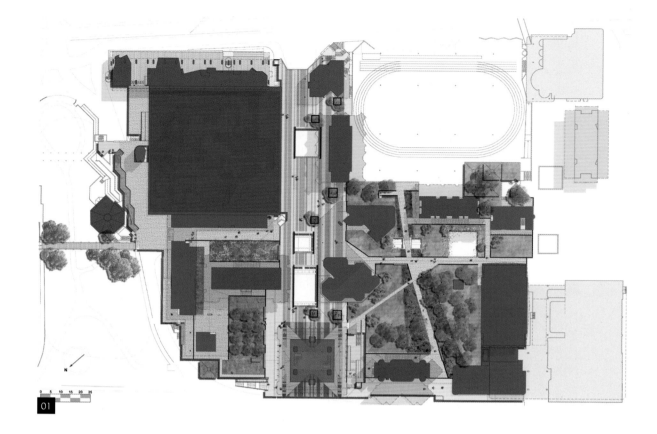

塞纳河岸街区
Front de Seine Neighbourhood

HYL

地点：法国巴黎
完工日期：2008
面积：2.3 ha
业主：SEMEA XV巴黎混合经济装备与开发公司
照片版权：Arnaud Yver (n°02, 05-09), Concepto (n°03, 04)

Location: Paris, France
Completion date: 2008
Area: 2.3 ha
Client: SEMEA XV
Photo credits: Arnaud Yver (n°02, 05-09), Concepto (n°03, 04)

01. 整体平面配置图，第一阶段工程
02. 主轴线上的散步道：Concepto公司的照明设计
03. 蓝色氛围：飘浮花园
04. 蓝色氛围：巴黎15区的空中广场

01. Master plan for the 1st phase of the works
02. Lighting design by Concepto: the main axis walk
03. Blue moods: the floating gardens
04. Blue moods: the esplanade of the 15th arrondissement

塞纳河岸街区的楼板平台式城市规划曾经是象征巴黎现代化的标志性工程，而目前则需要对其平台的防水系统进行整体翻修。HYL事务所首先参与了总体重新定位计划书的制定，随后完成了第一个阶段的施工。

这个翻新工程以三个准则为依据：打破街区与周围环境脱离的现状，使它向塞纳河和巴黎敞开；强化街区的识别特征；促使公众对街区楼板平台的使用。为此，首先要着手理清当前错综复杂的人行系统，通过设置新的垂直交通，使它与周边环境产生更好的连结。几条具有整合作用的重要轴线贯穿本街区而向外展开，犹如与周边街区的连结线。这些配置着楼梯和电梯的水平轴线引导着平台上新设公共空间的方位组织，以及新花园中小径的开设路线，花园里茂盛的植被为平台提供了全新的舒适氛围。另外，靠近塞纳河一侧重新建造平台边缘的护墙并在其上开洞，以塑造向河面伸展的视景。

我们重新恢复了1970年代为鸟瞰而设计、并与当时建筑风格相呼应的铺地图案。柔和而有趣味的照明设计使这项"重新带来活力"的方案更加趋近完美。

An emblem of Parisian modernism, the paving of the high-rise Front de Seine neighbourhood needed to be completely re-waterproofed. HYL first took part in drawing up an overall plan for updating the site, then undertook the first phase of the works.

Three main principles direct the renovation of the paving: breaking with its isolation by opening it up to Paris and to the Seine; strengthening its unique identity; and encouraging collective use of the paved area. First of all the architects had to clarify an almost labyrinthine network of paths and create better links between the Front de Seine and its urban environment through new vertical liaisons. Large unifying perspectives have been created throughout the neighbourhood to provide visual links with the rest of the 15th arrondissement. Punctuated by stairs and lifts, they direct the new public spaces of the paved area and guide the line of pathways through the new gardens, whose sometimes exuberant vegetation gives the paving a new grace. On the Seine side, the river wall has been rebuilt with openings to offer views down to the river.

The surface design harmonises with the architectural aesthetic of the 1970s, echoing the scale of the old paving motifs that were designed to be seen from the sky. Soft and playful lighting completes this "revitalising" approach.

05. 绿色小径
06. 一个崭新的生活环境
07. 住在花园里
08. 繁盛的植物向高楼进攻
09. 大楼脚下的野餐

05. Green walks
06. A new living environment
07. Living in a garden
08. The plant world confronts the towers
09. A picnic under the towers

03 城市公共空间 Urban public spaces

罗纳河岸
Banks of the Rhône
IN SITU

地点：法国里昂
完工日期：2007
面积：10 ha
业主：大里昂地区联合组织
合作设计师：Jourda architectes
照片版权：In Situ

Location: Lyon, France
Completion date: 2007
Area: 10 ha
Client: Le Grand Lyon
Co-project manager: Jourda architectes
Photo credits: In Situ

01. 自行车道和和本植物区
02. 罗纳河左岸，位于金首公园和杰尔朗公园之间的5千米长堤岸
03. 吉约蒂耶尔区段的浅水池

01. Cycle path and grassy islands
02. The left bank of the Rhône, 5 kilometres between the parks of Tête d'Or and Gerland
03. The Guillotière water wedge

在里昂，罗纳河左岸地区沿着河流5公里的长度伸展，覆盖了近乎10公顷的土地，其中包括了城市的中心区。里昂城市规划委员会决定采取措施把这片逐渐被停车场占据的河岸还给行人，并借此找回昔日分布其上的旧码头。In Situ事务所于2003年获得设计委托，对金首公园和杰尔朗公园之间的曲线河道周边进行规划，让一系列丰富的公共空间和自然空间交错设置，在河岸上延展开来。

这个位于河流主河床上的基地必然受到潮汐节奏的影响，设计师借鉴于大自然现象而发展出一个简单、开放和持久的整治方案。根据一个顺畅、流动的设计构图，几条不同的步行道和自行车道时而分开、变宽，又时而重新融合在一起，与罗纳河自身分汊型河道的形态相呼应。这组柔性交通路线伴随着河流延伸成一条漫长的散步空间。位于石砌护坡和法国梧桐脚下的下码头地区横断面宽度变化多，这些地形特征促使方案设计出特质各异的场所，在上游和下游地段是较为自然化的空间，中央地带则较为城市化。

这条公共路线同时是"公园和散步道"，它不仅伴随着河流伸展，也连接着沿岸的各种不同场所和生态环境。这也是一片提供多样使用功能和邻里交流的空间，从大城乡区域尺度和谐地过渡到它所穿过街区的尺度，巧妙地串连了各类场所：适用于大型集会、游戏、餐饮、运动的空间，以及沿着水边漫游路线而设置的较为隐密的休憩场地。这个既富有弹性又具有持久性的规划方案把时间因素纳入考量，因此也同时提出了将来对不同空间进行维护和管理的策略。

In Lyon, the banks on the right side of the Rhône cover almost 10 hectares in a 5 kilometres stretch at the heart of the city. Lyon's city authorities have committed to the reconquest of this site in order to give these old ports that have been progressively confiscated by car parking back to pedestrians. After the definition studies submitted in 2003, In Situ developed a continuous project that will produce a fertile furrow of public and natural space in the bend in the river between Tête d'Or and Gerland parks.

In the flood plain of the river, this land area is subject to the rhythm of high waters. This inspired a development that had to be simple, open and strong. Thus the different pedestrian and cycle paths diverge, widen or come together like the "interlacing" phenomenon of the Rhône itself, through a fluid and dynamic design. This network of soft transport modes stretches out in a long promenade beside the river. At the foot of the long stone embankment and the plane tree mall, the Low Ports space offers very varied cross-widths. This arrangement leads to strong contrasts, from the most natural places up and down river to the most urban in the central part.

Both a "park" and a "promenade", it is a public way that links several places and environments and follows the river. It is also a land area with different uses in contact with each other that reconciles scales between the urban area and the neighbourhoods it crosses, and between the spaces for events, for play, for eating, for sport and restful, more private places for walking beside the water. This project, which is simultaneously supple and sustainable, also integrates a maintenance and management strategy for the different spaces that takes account of time.

04. 河岸森林
05. 花园岛和"条板散步道"
06. 上游布雷蒂洛区段和河岸森林的场景图

04. Riverbank environment
05. The garden islands and "the boardwalk"
06. Scenographic view of Brétillot and the upstream riverbank environment

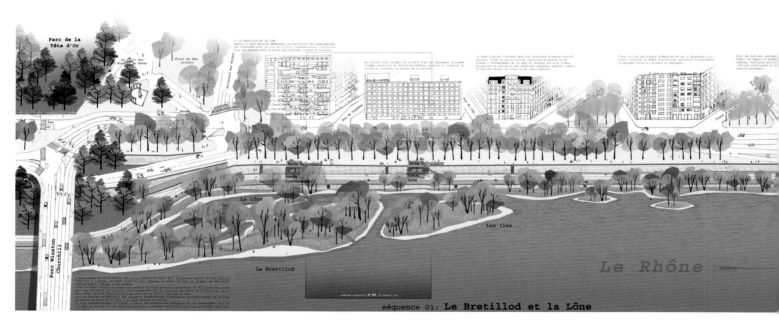

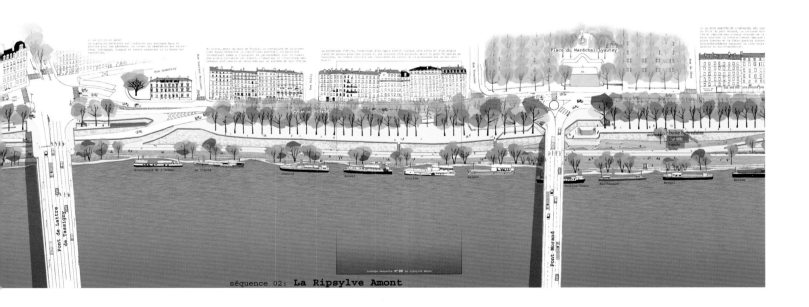

séquence 02: **La Ripsylve Amont**

03 城市公共空间 Urban public spaces

07-09. 游戏场和休憩躺椅
10. 天桥和游戏区剖面图
07-09. The playgrounds and sunloungers
10. Section of the footbridge and slides

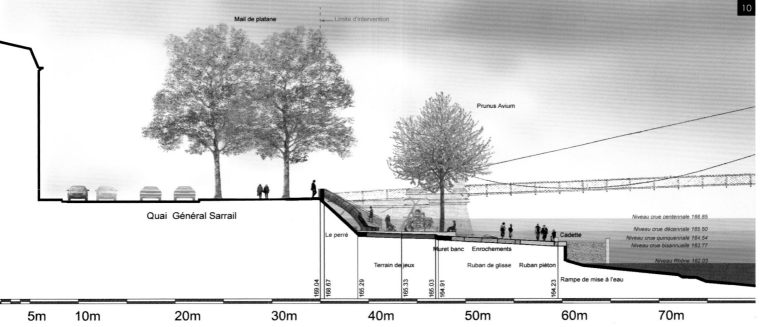

03 城市公共空间 Urban public spaces

11-13. 罗纳河大草坪
14. 大草坪与条板散步道区段的场景图

11-13. The grand meadow of the Rhône
14. Scenographic view: grand meadow and the boardwalk

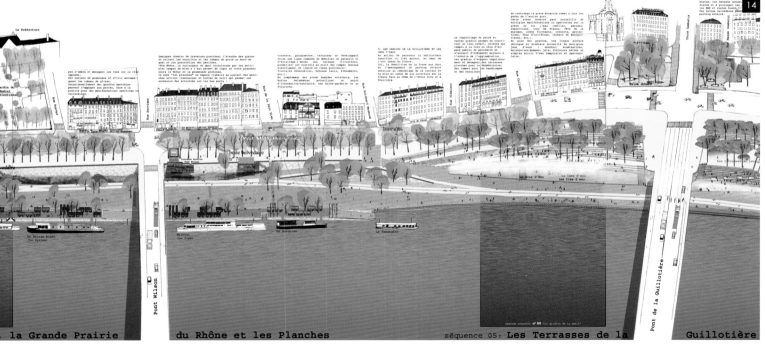

03 城市公共空间 Urban public spaces

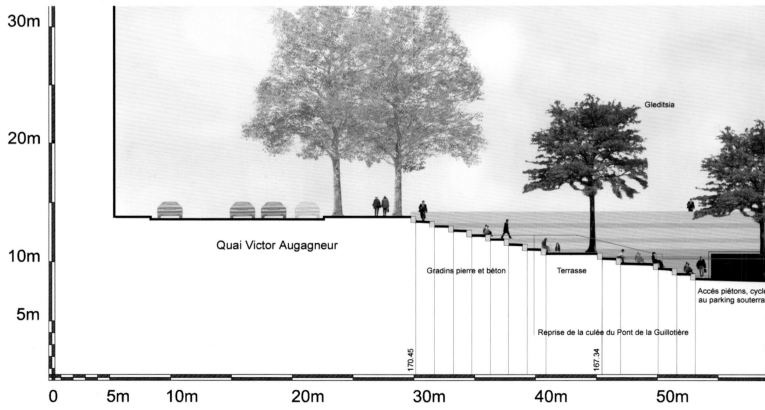

15、16. 吉约蒂耶尔区段的阶梯平台和浅水池
17. 吉约蒂耶尔区段的阶梯平台和浅水池剖面图
15-16. The Guillotière terraces and the water wedge
17. Cross-section of the Guillotière steps and the water wedge

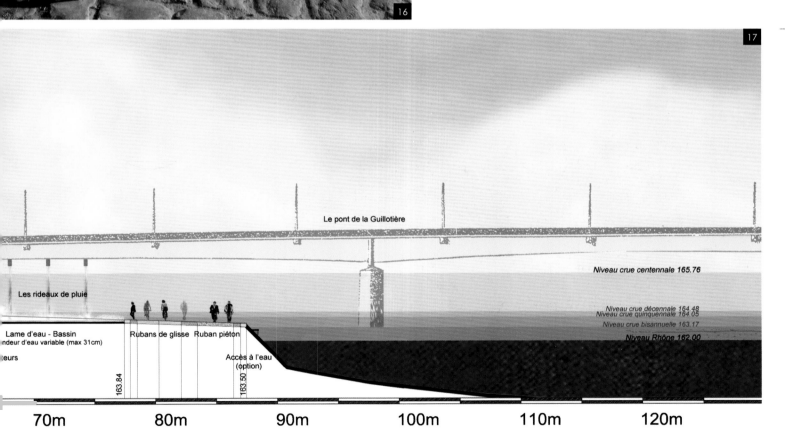

03 城市公共空间 Urban public spaces

18. 面对主宫医院和富尔维耶教堂的河岸景观
19、21. 吉约蒂耶尔区段的阶梯平台
20. 碗状溜冰场和稍远处的罗纳河游泳池
18. View of the Hôtel Dieu and the Fourvière basilica
19&21. The Guillotière terraces
20. The skateboard bowls and the Rhône swimming pool in the background

03 城市公共空间 Urban public spaces

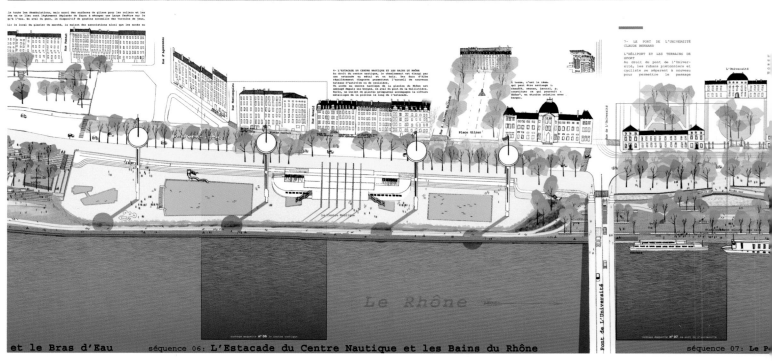

séquence 06: L'Estacade du Centre Nautique et les Bains du Rhône

22. 罗纳河游泳池堤岸步道
23. 在树荫下的滚球场
24. 平台广场和大学桥
25. 平台广场和大学桥区段的场景图

22. The boom of the Rhône swimming pool
23. The boules pitches under the trees
24. The esplanade and the Université bridge
25. Scenographic view of the boom and the Université bridge

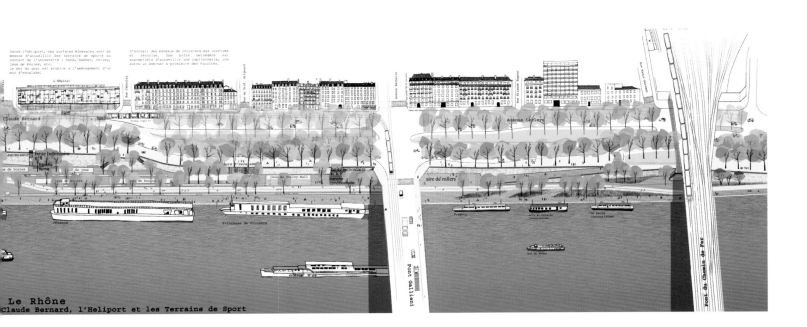

03 城市公共空间 Urban public spaces

26. 下游的河岸森林和河流植物廊道
27. 河岸的夕阳西下
28. 河流植物廊道区段的场景图

26. The downstream riverside environment and the river botanical gallery
27. Sunset on the banks
28. Scenographic view of the river botanical gallery

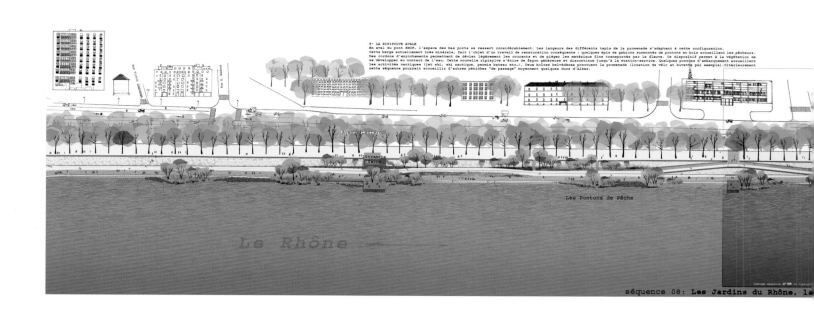

03 城市公共空间 Urban public spaces

Pajol ZAC
帕若尔协商开发区
IN SITU

地点：法国巴黎
完工日期：2013
面积：9 500 m²
业主：巴黎市政府
设计总负责：Jourda architectes
图片版权：In Situ

Location: Paris, France
Completion date: 2013
Area: 9 500 m²
Client: Paris City Council
Project manager: Jourda architectes
Image credits: In Situ

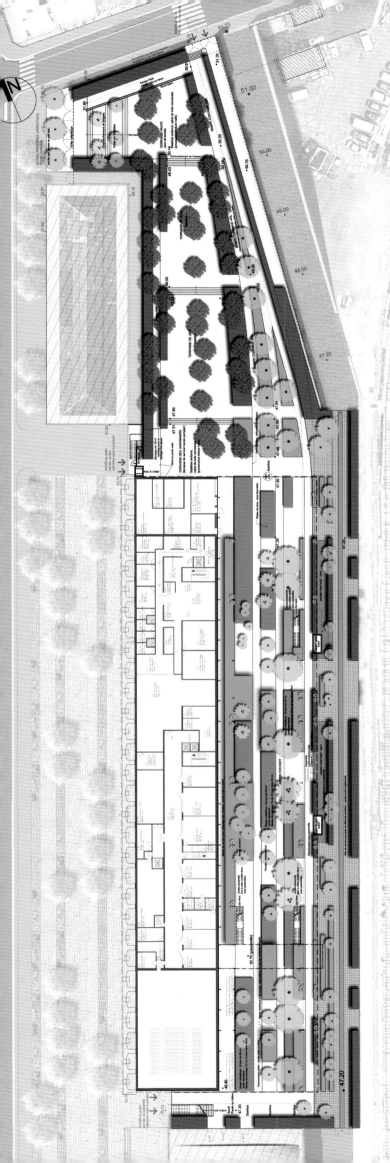

01. 帕若尔协商开发区的前庭广场
02. 整体平面配置图

01. Forecourt of Pajol ZAC (comprehensive development zone)
02. Master plan

沿着巴黎东站的铁轨用地，帕若尔花园在一片广阔基地上从北向南延伸着。这个花园通过两个序列空间带人展开一段连续的旅程：被保留并经由建筑师F.H茱兹改造的昔日帕若尔市集大厅容纳着一片室内花园，而露天花园则向北延伸与街区的城市广场衔接在一起。这片线性花园伴随着或平行或者交错的铁道线，在一个由植物墙篱组成的网络上，以各种长条状元素来组织空间。

露天花园在帕若尔厅的北侧展开，其错落有致的平台与地形巧妙地结合，并设置着儿童游戏场、一个集会广场和若干开放的草坪，成为充满活力的公共空间。每个平台上皆点缀着常青松树丛。在基地南侧的室内花园，即阴凉花园和白色花园，从室外轻轻滑入市集大厅的半边屋檐下，并为大厅另一边改建成的青年旅馆带来光线。在市场大厅广阔顶棚的保护和遮阴下，人行小径循着旧轨道遗迹伸展，构成一片植被密集的空间。在市集大厅和沿着铁道的小径之间，则展开了一系列共享式花园，形成一道长长绿带。

Along the immense land area of the Gare de l'Est railway lines, the Pajol gardens extend north-south at the widest part of the site. This long garden describes a continuous path in two sequences: under the Pajol covered market hall, which has been preserved and rehabilitated by F.H. Jourda, is a covered garden, while the open-air garden unfolds to the north and meets the neighbourhood's urban esplanade. This linear space follows the mesh of parallel or converging lines of the railway tracks, and is organised in successive strips in a network of planted corridors.

The open-air garden unfolds north of the market hall. Its stacked terraces, which make best use of the topography, host children's playgrounds, an events space and free lawns to make up a dynamic public space. A series of pine groves punctuate each of these clearings. In the south, the covered garden, shade garden and white garden are under the market hall and filter the light of a wide border, which the youth hostel gives onto. The pedestrian walks follow the lines of the railway to compose a space that is densely planted, protected and shaded by the cover of the market hall. Allotments are organised in a long strip between the market hall and the exterior walk along the railway tracks.

03. 阴凉花园以林下植物作为主要植被种类
04. 边缘与空地，方案所选择的植被塑造出过滤效果
05. 小径的图案以铁道线条作为参考
03. Underwood species for a shade garden
04. Borders and clearings, the planting creates a filter
05. The design of the avenues is based on the old railway lines

雨水的回收让一个特殊系统得以在花园中实现：安置在旧轨道之间的几个水池接纳并且储存屋顶落下的雨水，然后通过自动灌溉系统再把雨水输送给所有花园。某些水池中种植了水生植物，另外一些水池则变成平滑的镜面。这些火车站旁的线状花园为周边街区的居民提供了一处幽静的散步空间。雨水从屋檐落下的滴答声、清新自然的氛围使其成为巴黎市内一处环境优美的室外空间。

Rainwater harvesting allows for a specific system to be put in place: several basins planted between the old railway tracks harvest and stock rainwater from the roof, which is used for the automatic watering of all the planted areas. Some of these basins are planted with an aquatic vegetation, while others form ornamental pools. These linear railway gardens on the fringes of the station offer a calm place to walk for the inhabitants of the neighbourhood. The lapping of water from the rain curtains and the freshness of the gardens make it a special place in Paris.

03 城市公共空间 Urban public spaces

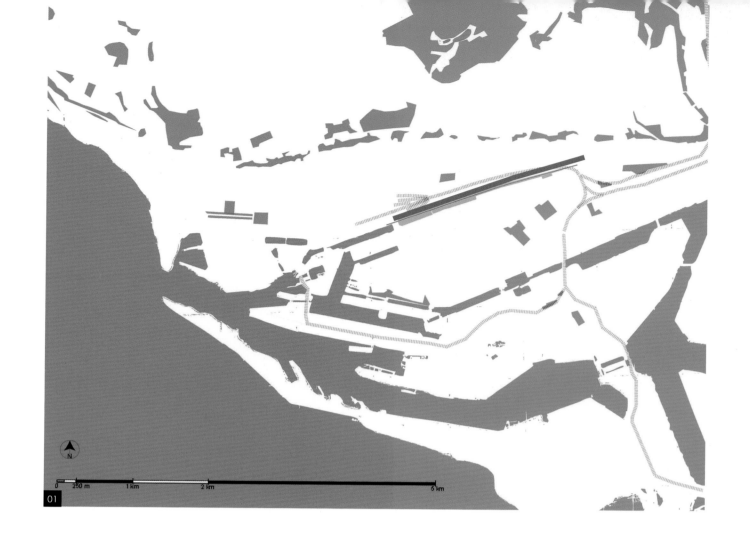

勒阿弗尔城市入口
Le Havre City Entrance
L'ANTON & ASSOCIÉS

地点：法国勒阿弗尔
完工日期：2007-2018
面积：20 ha
业主：勒阿弗尔市政府
图片版权：Agence L'Anton & Associés

Location: Le Havre, France
Completion date: 2007-2018
Area: 20 ha
Client: Le Havre Town Council
Image credits: Agence L'Anton & Associés

01. 基地位置图
02、03. 鸟瞰全景：现状与方案
01. Location plan
02-03. Existing and projected overall views

对勒阿弗尔市来说，丘吉尔和列宁格勒林荫大道的整治计划是一个相当重要的挑战。这条近2.5千米长的道路连接了港口、南部街区和市中心。然而此条道路呈现出完全属于机动车的面貌，位于交叉路口下方供汽车穿越的小型地下通道、高架桥甚至立交桥一个接着一个。道路沿途经过城市中一些不断变化的街区，并为在整个城乡区域占有结构作用的若干发展中心提供服务。其道路空间组织(横向剖面)整体来说十分均匀，但是北段设有慢车道(停靠车道)，南段则设有反向车道。在反向车道和林荫大道之间，将近1千米长、50米宽的隔离岛被三个加油站、低使用率的停车场和若干奄奄一息的活动设施所占据。

朗东景观事务所提出一个大幅度利用现有设施的整治方案，在北侧的慢车道重新种植树木和改善空间质量。标准路段设置了双向四车道。在南侧车道回收的面积范围内则开辟了一条步行道和自行车道。此外，交叉路口被重新复原为平面交叉、小型地下车道因而被取消。同时在靠近"沃邦码头"的一座桥也被拆除，以便提升城市中心入口和周围新建设施的意象。一个新的带状公园被设置在原来三个加油站用地上，不仅提升正在进行的规划项目的价值，也为勒阿弗尔市南部的街区提供近5公顷的全新绿色空间，并且建立了雨水储存系统和去除城市地下潜水层污染的系统。

The redevelopment of Churchill and Leningrad Boulevards presents a major challenge for Le Havre. Over almost 2.5 kilometres, this road, which serves the port, the southern neighbourhoods and the town centre, has the image a trunk road where traffic tunnels under the junctions, flyovers and uneven interchanges follow on in quick succession. The road runs alongside sectors that are ripe for change and serves the development clusters that will structure the urban area. The transverse profile of Churchill and Leningrad Boulevards is relatively homogenous with the service road to the north and the road running in the opposite direction to the south. Between the latter and the central lanes a traffic island almost 1 kilometre long and about 50 metres wide is occupied by three petrol stations, parking and a few activities in decline.

The proposal of L'Anton & Associés re-uses what is there as much as possible. In the north, the service roads are improved and planted with new rows of trees. The road surface is brought down to 2x2 lanes in the traffic section, to the benefit of a pedestrian and cycle path. The junctions have been levelled and the tunnels closed. A bridge at the "Vauban Docks" has been demolished, improving the approach to the city centre and the facilities already in existence or to come. The service stations have been replaced by a park, which presages the urban projects still under construction and offers almost 5 hectares of green space for the southern neighbourhoods of Le Havre. It also provides a system of stormwater harvesting and soft remediation of the first flush urban runoff.

04. 标准路段双向四车道剖面与平面图
05. 交叉路口剖面与平面图
06-07. 自行车与行人使用的散步道
08. 城市小品与照明系列
09. 抵达市中心

04. Typical section and plan of the 2x2 lanes highway
05. Typical section and plan of a junction
06-07. Pedestrian and cycle paths
08. Furniture and lighting designs
09. Arriving at the town centre

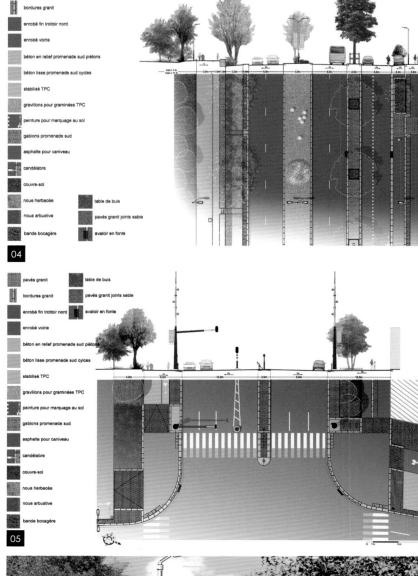

这个方案在保持道路主要特征的前提下简化了其功能与运作方式，因此能够将自然空间引入城市，创造出一系列的散步道、一个由三个主题花园组成的线形公园、一个观景平台以及一个大型的芦苇丛园。重新为城市营造出港湾氛围，让河流、芦苇和城市空间能够产生对话。

This redevelopment simplifies the functioning of the road while making the most of its main characteristics. A place for nature in the city can now be installed through the creation of walks, a linear park made up of small themed squares, a belvedere and a huge reed-bed park. This project thus gives the city back an estuary ambiance, a meeting place between the river, its reeds and the urban space.

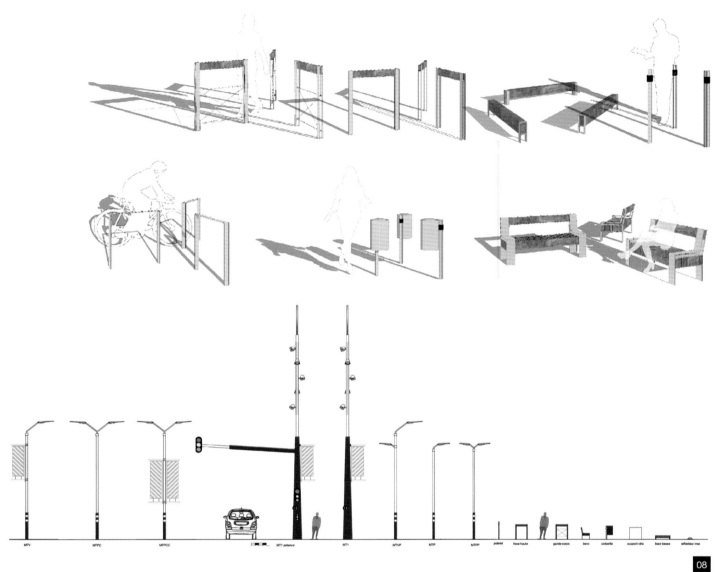

03 城市公共空间 Urban public spaces

10. 未来的码头公园
11. 剖面图：城市入口轴线、码头公园和平行道路
12. 方案整体平面配置图和工程进展日期
13. 在码头公园散步

10. The future Docks park
11. Section of the town entrance axis, the Docks park and the parallel roads
12. Master plan of the project and time scheme for the construction work
13. Walking in the Docks park

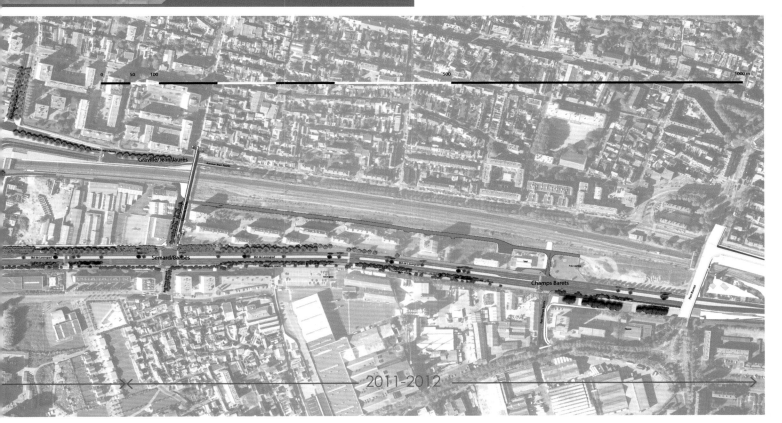

03 城市公共空间 Urban public spaces

14. 旧市场建筑转变为城市广场的三个阶段
15. 旧市场建筑的灯光设计
16. 旧市场建筑附近的自行车道和人行步道
17. 收集雨水的绿化斜沟
18. 金仙草、鸢尾花、桦树在散步道旁绽放色彩

14. The three stages of the total rehabilitation of the old market hall into an urban square
15. Lighting the market hall
16. Pedestrian promenade next to the market hall
17. Stormwater harvesting swale
18. Purple loosestrife, iris, birches and other trees colour the borders of the walks

03 城市公共空间 Urban public spaces

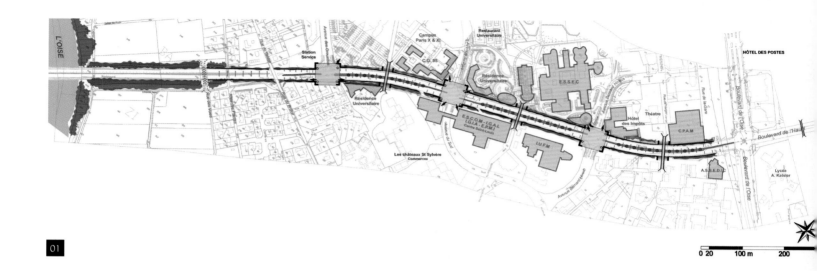

01

奥蒂埃林荫大道
Hautil Boulevard

L'ANTON & ASSOCIÉS

地点：法国塞尔日-蓬图瓦兹
完工日期：2010-2011
面积：3 ha
业主：塞尔日-蓬图瓦兹城乡区域联合组织
照片版权：Agence L'Anton & Associés

Location: Cergy-Pontoise, France
Completion date: 2010-2011
Area: 3 ha
Client: Communauté d'Agglomération de Cergy-Pontoise
Photo credits: Agence L'Anton & Associés

01. 整体平面配置图
02. 从一个天桥上看到的方案景观
03. 新设置的机构从此由这条重新整治过的大道直接通达

01. Master plan
02. Overall view from a footbridge
03. The new establishments are now accessible from the boulevard

塞尔日-蓬图瓦兹市的奥蒂埃林荫大道是通向省政府中心街区的入口，也是这个地区的主要交通轴线之一。此大道的整治被包含在一个以加强城乡凝聚力为目标的综合计划当中："城市核心区"项目、ESSEC高等经济商业学院的扩建、公园和省政府周边的整治等。这条道路必须成为其他主干道的重新定位的参考模范。

这条林荫大道建于1971年，当时的城市规划设计出高低变化的道路并且把人行交通布置在机动车系统之外（设置过街天桥）。施工过程持续了许多年，居民们必须穿越各个不同的工地，或者透过步行通道来抵达市中心的所有服务设施：商业、学校、行政机构等。因此，今天的林荫大道呈现出完全被机动车占据的面貌，其上的城市小品和设施网络则陈旧不堪。然而，新近竣工的建筑开始出现在道路沿线，公共汽车停靠站的使用也越来越频繁，这些改变开始把更多的步行功能带到这片还未经改造的空间中。

One of the main traffic arteries of Cergy-Pontoise, Hautil Boulevard forms the entrance to the Grand Centre Préfecture neighbourhood. Its redevelopment is part of a larger project for strengthening the centrality of the urban area: the operation "city heart", an extension of the ESSEC business school, the reconfiguration of the park and the surroundings of the prefecture, etc. This boulevard has to serve as a reference for the improvement of other structuring roads in the urban area.

The urban project of 1971 led to the staggering of levels on the boulevard and encouraged pedestrian circulation outside the main traffic arteries as the town was going to be under construction for several years. The new inhabitants certainly had to cross various building sites, but could also access all the services of the town centre – shops, schools, administrative buildings – on foot via clean spaces. As a result the boulevard today looks like a trunk road and its urban furniture and networks are at the end of their life. The most recent buildings have started to go up along the boulevard, and the bus stops, with an increasing number of users, force pedestrians onto areas not designed for this type of use.

04. 标准路段和交叉口的平面配置原则
05. 标准路段平面和剖面图
06、07. 交叉路口边缘的小广场
08. 居高临下的"历史"城区和大道上的新设施机构

04. Principle master plan, road section and junction
05. Plan-section of the road
06-07. Small squares created next to the junctions
08. The "historic" town looking down and the new establishments at street level

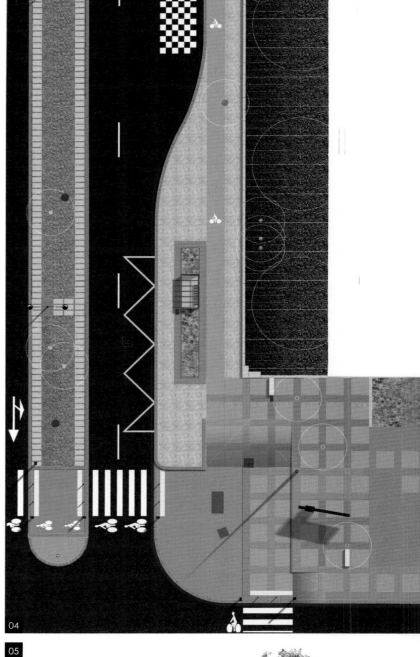

这个项目开展了非常深入的公众协商过程，以便重新为大道的功能进行定位，终而决定在大道上开辟公共汽车专用道、自行车道和舒适的人行道，以更为城市化的方式来进行绿化，在道路交叉口开辟利于交流的空间，并且设置更符合需求的照明系统。基于明智抉择与经济节约的理由，一些剩余的空间被最大化地利用（中央隔离岛、道路面、水渠、斜坡），标准路段的路面照明被减低，以优先赋予人行道、自行车道和交流空间较为良好的灯光设施。通过质量的提升、柔性空间尺度的加大以及交叉路口的分级处理，此大道从此拥有截然不同的新面貌。

An in-depth consultation was carried out to redefine the functioning of the road: integrating dedicated bus lanes, cycle lanes, pavements that are comfortable to walk on, more urban plantations, places of conviviality near junctions and a suitable lighting plan. For reasons of good sense and economy, elements that were already there have been used as much as possible (a central island, road surfaces, water courses, talus). The lighting of the pavements, the cycle paths and the crossroads have been prioritised over the road surfaces used by traffic. The boulevard has acquired a more urban image through qualitative treatments, but also through the generous dimensions of the "soft" spaces, and by a differentiated treatment of the junctions.

03 城市公共空间 Urban public spaces

09、10. 软性交通（人行道、自行车道……）
11-13. 让植物的蓬勃活力凸显出来，并组织它们的发展可能性
09-10. Pedestrian and cycle paths
11-13. Designing the plant dynamics and organising the use of the land

03 城市公共空间 Urban public spaces

Mazelle Square
马泽尔广场
FLORENCE MERCIER PAYSAGISTE

地点：法国梅斯
完工日期：2011
面积：3.5 ha
业主：梅斯市政府
照片版权：Florence Mercier (n°01, 05-07), Marc Petit Jean (n°08)

Location: Metz, France
Completion date: 2011
Area: 3.5 ha
Client: Metz City Council
Photo credits: Florence Mercier (n°01, 05-07), Marc Petit Jean (n°08)

01. 以现代化的景观语汇来作为老城市和新街区之间的空间转换
02. 平面配置图：一个位于艺术城市与水文城市交汇点的广场
03. 塞尔河畔的散步道
04. 为城市居民规划的景观空间

01. A contemporary vocabulary brings the old town together with recent neighbourhoods
02. A square at the crossroads between the city of art and the city of water
03. The Seille riverside walk
04. An urban scene that's all about the citizens

马泽尔广场位于梅斯的老城区与新蓬皮杜中心所在的南部新城区的交界处。从城市发展的关键性和文化活力的层面考量，马泽尔广场所在街区是极具策略优势的位置。从前，这里一直是塞尔河几条支流的流经之地，如今则处于被文化浪潮所冲击的交叉路口。因此，方案通过在这个艺术和水相会的城市中建立一个广场，以隐喻的手法表达了大自然在城市中心的存在。

广场由一个硬地面的大型开放空间所组成，为建筑立面提供了一个基座，也成为设置露天咖啡座的场地。在南边，叶子稀疏的树木让人联想到河湾植被，它们被种植在设有喷泉的几个硬地面空间的周围，形成了更私密的空间。

作为城市的标志，此广场也同时是城市献给居民和艺术的大舞台。在花园中心，一些雕塑活跃了空间的气氛，与行人的路线相互交错。夜晚，高大的灯杆形成了戏剧性的照明效果，与铁路对面的蓬皮杜现代艺术中心遥相呼应。

At the point where the old city of Metz meets the more recent neighbourhoods in the south, which notably include the new Pompidou Centre, Mazelle square is strategically placed in terms of urban challenges and cultural dynamism. Occupied in the past by the different branches of the River Seille, and today at the crossroads of cultural initiatives, the project expresses the metaphor of nature at the heart of the city by offering a square where the city of art meets the city of water.

The square is composed of a large, open, hard surface that provides a sold base for the facades of the buildings around it and hosts café terraces. In the south, trees with light foliage evoke the vegetation of river meanders; in their shade more intimate corners are arranged around stone slabs with water flowing over them.

A figurehead for the city, the square is also an urban scene dedicated to the citizens and to art. At the heart of the gardens, sculptures enliven the space and play with the shapes made by people as they pass by. At night, tall masts create a scenic lighting that enlivens the square like an echo of the Pompidou Centre situated on the other side of the railway.

05、06. 水柱喷泉的节奏和椅凳的线条
07. 广场将城市延伸出来，而机动车道被改造为林荫道
08. 充满节庆氛围的空间
09. 夜间景观

05-06. Fountain rhythms and bench lines
07. A square that extends the city centre, and roads reworked to form a boulevard
08. A theatrical space for cultural events
09. A nighttime scene

03 城市公共空间 Urban public spaces

弗朗索瓦·密特朗林荫道
François Mitterrand Mall
MUTABILIS

地点：法国雷恩
完工日期：2013
面积：宽度50 m，长度约为750 m
业主：雷恩市政府
图片版权：Mutabilis

Location: Rennes, France
Completion date: 2013
Area: 50 m in width, about 750 m in length
Client: Rennes City Council
Image credits: Mutabilis

01. 以行人为优先的林荫道
02. 整体平面配置图：位于伊勒河和维兰河之间的方案

01. The mall, a space where pedestrians are prioritised
02. Master plan of the project between the Ille and the Vilaine

弗朗索瓦·密特朗林荫道是雷恩市的历史轴线。它位于一处低洼地，历史上是雷恩人散步和交往的场所，今天的林荫道则是一条结构性的交通轴线，成为城市入口和城市西部的交通转换场所，同时也是一个代表性空间和大型集会的场所。然而，它已经逐渐被机动车所侵占，机动车不仅逐渐占据了整个中心区域，也使其转变为一个单调、不利于步行穿越的空间。此林荫道同时处于一个在不停变化中的地段，在这里，很多地块、建筑和相邻街区被挑选出来作为旗舰项目（例如：布列塔尼广场、拉马比来协议发展区、林荫草坪、圣西尔河堤、维凯尔广场以及由让·努维尔设计、即将建成的卡普林荫大厦）。

此项目的重点在于为弗朗索瓦·密特朗林荫道重新定位和进行改造整治，目的是为行人建立一个宽敞的步行空间，强化它的绿化轴网，展现其适合各年龄人群、具有多用途的强烈特征。此整治方案也为现有的服务业和商业（特别是餐馆）提供了崭新的展示橱窗。

方案在沿着林荫道北立面的一侧优先设置了宽阔的行人空间，重新赋予了树木伸展的空间，并设计了一个具有现代风格且趣味盎然的铺地构图。此构图激发步行者的好奇心，并引导他们来到合流花园，在此眺望伊勒河和维兰河的交汇。

François Mitterrand Mall is a historic axis in the city of Rennes. Built on marshes, it was the traditional place for the city's inhabitants to stroll and meet each other. Today, the mall is a structuring road that forms an entrance route to the city and a transit route from the west of Rennes, as well as a place for municipal events and gatherings. It has been progressively invaded by cars, which occupy its entire centre, and has become a monotone area that is no longer pleasant to walk along. At the same time, it forms part of a changing sector whose parcels, buildings and adjacent neighbourhoods have been singled out for flagship projects (Bretagne Square, the ZACs – comprehensive development zones – of La Mabilais and Pré du Mail, Saint Cyr quay, Vicaire Square…, and the future Cap Mail building designed by Jean Nouvel).

The development project for François Mitterrand Mall aims to make it a welcoming space dedicated to pedestrians, enhancing and strengthening its network of planted areas and asserting a strong identity so as to encourage various uses that allow the generations to mix. It is also a means to offer the existing services and businesses, particularly restaurants, a new and more attractive setting.

Pedestrians take priority on the wide quayside attached to the north facade, where the trees now have space to breathe and a playful, contemporary paving has been laid. The design of the mall will pique people's curiosity and guide walkers to the Confluence Garden, which overlooks the point at which the Ille meets the Vilaine.

03. 城市座椅和漆绘了图案的地面
04. 林荫道北侧与建筑立面产生紧密的关联
05. 此林荫道是为行人设计的空间
06. 向维兰河敞开的合流花园
07. 弗朗索瓦·密特朗林荫道的横向剖面图

03. Seats and painted plaza
04. The north quay clings to the façade
05. The mall, a space with many uses dedicated to pedestrians
06. The Confluence garden, opening onto the Vilaine
07. Cross-section of François Mitterrand Mall

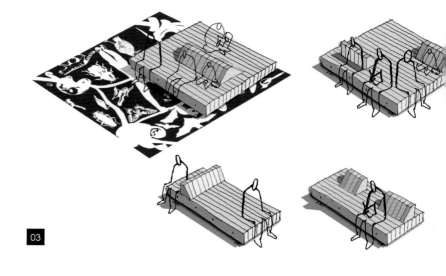

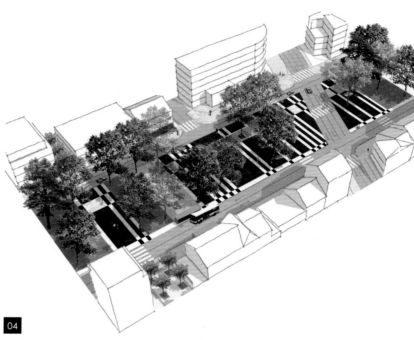

03 城市公共空间 Urban public spaces

四季广场 *Seasons Square*

OLM / PHILIPPE COIGNET

地点：法国拉德芳斯	Location: La Défense, France
完工日期：2011	Completion date: 2011
面积：8 000 m²	Area: 8 000 m²
业主：EPADESA	Client: EPADESA
合作设计师：David Serero Architecte Urbaniste (设计总负责)	Co-project manager: David Serero Architecte Urbaniste (project representive)
照片版权：Jean-Marc Charles (n°04), OLM (n°01-03, 05-09)	Photo credits: Jean-Marc Charles (n°04), OLM (n°01-03, 05-09)

01. 设计竞赛阶段透视图,从第一大厦俯视广场
02. 地下停车场出口和红叶枫树
03. 通风格栅完全融入木板铺地
04. 广场以LED(发光二极管)和金卤灯进行夜间照明

01. Perspective drawing for competition, view from the First Tower
02. Parking payment kiosk and red maples
03. Ventilation grille integrated into the wooden decking
04. LED and metallic iodide lighting

拉德芳斯一区是巴黎从1960年开始进行的主要城市改造的最后几个项目之一。借助于未来艾尔米塔日建筑群建成的机会,当下福斯特大厦(原AXA保险大厦)的改造和四季广场的整治使得此街区得以逐渐转向塞纳河。

方案通过清除所有显露在广场上的技术设备、水泥花台、零散种植的植物、几十年来积累下来的样式各异的围栏和铺地,重新强化了这个公共广场的中心角色。景观师在此建立一个灰色混凝土地面,以便整合广场和周围相邻的建筑,它将随着相邻建筑的改造进程,逐渐展开。方案在整个空间的中央和改造后的地下停车场的入口对面,设置了一个木质铺地空间,其上设置着由枫树和李树组成的绿化结构和供人们小憩的座椅。这些座椅上安装了LED(发光二极管)光源,使整个广场散发出蓝色微光。广场原有的一群桦树被保留了下来,使绿化网络更加完整,并赋予广场一种植物的季节性,完全符合了广场名称的意义。

本方案因而得以探讨几个重要课题:建立建筑立面和公共空间之间的关系,以步行作为空间的主导功能,以及一个公共广场在城市建设中应具有的识别特征。

The neighbourhood of La Défense 1 is one of the last to undergo major transformations since its creation in 1960. The rehabilitation of the First Tower (the former AXA tower) and the redevelopment of Seasons Square inaugurate a neighbourhood that will eventually face the Seine, thanks to the construction of the future Hermitage towers.

The project reaffirms the central and intersecting role of this public space by removing all the technical equipment from sight, as well as the concrete planters, the widely spaced plantations, the collections of railings and paving accumulated over several decades. The plan is to create a plinth of grey concrete that unifies the square and the neighbouring buildings, which will spread as the rehabilitation of the buildings progresses. In the centre of the space and facing the rehabilitated car park entrance it contains a second plinth in wood, which is both the setting for a new planting of trees, composed of maples and cherry trees, and of benches for sitting or reclining. The benches integrate LED lighting, which diffuses a blue light across the whole square. The original birches have been kept and complete the tree network to give Seasons Square, as its name suggests, plants for all seasons.

The project thus questions the relationship between facades and public space, the role of pedestrian circulation as the square's function and the basis for its design, as well as notions of identity in the urban fabric of a public square.

05. 地下停车场出口的玻璃反射
06. 橡木长凳和铺板
07. 椅凳和种在木箱里的枫树
08. 基地原本即存在的桦树
09. 天人菊花圃

05. Reflections of the payment parking kiosk
06. Decking and benches in oak wood
07. Benches, maples in the planters
08. The existing birch trees
09. Gaillardia beds

03 城市公共空间 Urban public spaces

帕甬河畔景观步道
Paillon Landscape Walks
PÉNA & PÉÑA PAYSAGISTES

地点：法国尼斯
完工日期：2013
面积：10 ha
业主：尼斯-蔚蓝海岸城市联合组织
图片版权：Christine & Michel Péna (n°04-08), Golem & Péna (n°01-03)

Location: Nice, France
Completion date: 2013
Area: 10 ha
Client: Communauté urbaine Nice Côte d'Azur
Image credits: Christine & Michel Péna (n°04-08), Golem & Péna (n°01-03)

01. 方案全景透视图：位于尼斯市中心的12公顷绿地，将剧场和海洋连接起来
02. 有如镜面的大片浅水池反射着四周景观的投影，它可以完全消失以便将空间让给大型活动使用
03. 抵达海岸的散步道以经过重整的阿尔贝一世公园
04. 黑白相间的铺地，越接近教堂，白色的铺地板块就越多

01. Cavalier perspective: 12 hectares in the heart of Nice, linking the theatre to the sea
02. The strip of shallow water that acts as an ornamental lake can disappear for major events to take place
03. Arriving at the sea through the Albert 1st garden, which has been redesigned and made more airy
04. The white slabs increase as you approach the church

位于尼斯的帕甬河是昔日妇人聚集洗衣的河流，承载着历史记忆。作为老村落的自然防线，帕甬河以性感的曲线蜿蜒绕过夏多丘陵（城堡丘陵）。然而，这份柔美却被周遭众多水泥建物所破坏，若要重拾这段美丽历史，则必须建立一条通往大海的绿化河谷，让人们重新认识尼斯最美的景致。

建设一条将文化街区联通到英国人漫步大道和海洋的1千米长"天然"绿廊成为方案的主要构思，在城市尺度上和在一般居民或观光客尺度上都能提供相当丰富的景观空间。为此必须拆除大型水泥建筑物以建设一系列花园，同时设置一条形态柔和、缎带般的石板大道。石板带最小宽度8米，能够穿越树木与各式障碍物，其上配置极具开放性的绿色与蓝色地毯，向四处旷景展开视野。这个通廊以景观散步道的形式赋予帕甬河整体空间一个新的组织秩序，使人们得以重新欣赏沿河美景。尼斯如画般的自然景观也呈现眼前，例如远处阿尔卑斯山的前锋、夏多丘陵以及与城市对话的海洋。

Nice remembers the Paillon, the river of the washerwomen. As the natural enclosure of the old village, it circumvented the Château Hill in a sensual curve. But this softness was martyred to so much concrete that it was necessary to rewrite this beautiful story by recreating a green valley leading to the sea, and revealing once again one of Nice's most beautiful landscapes.

Opening the "natural" link that leads from the cultural cluster to the Promenade des Anglais and the sea (1 km²) appeared as the fundamental ambition, capable of giving something substantial back both on the scale of the city and on that of the ordinary inhabitants of Nice its tourists. The large concrete buildings had to be demolished in order to create the gardens. A ribbon of stone with a fluid form was drawn, with a minimum width of 8 metres, which slides between the trees and the various obstacles. It runs alongside a wide and very open carpet of green and blue, offering sweeping views of Nice. The combination of these two elements allowed the landscape architects to make the place readable again, to reorganise all the spaces of the Paillon into a wide and attractive landscaped walk. It is now possible to admire Nice's most beautiful natural canvases, the faraway foothills of the Alps, the Château Hill and finally the sea revisited by the city.

05. 阶梯颜色黑白交替有如钢琴琴键，延伸了广场的铺地逻辑
06. 望向誓愿教堂的景观：地面高度的差异使人感觉不到道路的横向穿越
07. 石块铺地的原则同样延伸在阶梯上
08. 柑橘园的氛围弥漫在整个广场上

05. Steps like a piano keyboard continue the design of the contrasting stone slabs
06. Towards the church of the Vow: the difference in level obscures the road and prevents it from spoiling the view
07. The steps continue the carpet of stone
08. The original orange grove ambiance is extended over the whole square

此绿廊规划的第一阶段涉及誓愿教堂前庭和拉布尔嘎达广场的整治，景观师在此采用尼斯地区最常见的石材：石灰岩与玄武岩，以图片像素般的拼排方式（采取30x60cm的石板模块）来进行铺地，随机交错两种石材，从教堂旁边以白色主导的地面逐渐过渡到嘎利巴尔迪广场以黑色主导的地面。植被的规律系统也受到这个主要结构的影响，在靠近教堂的一侧布置得较为紧密，在另一侧玄武石铺面的广场上则分布得较为疏散。两种元素的交错的结果使基地的明亮度得以取得平衡。

To create the Bourgada esplanade and the forecourt of the church of the Vow, which form the first section of the greenway, limestone and basalt were chosen to continue a theme that had already proved itself in its use all over the city. The paving is laid like a pixelated image (the stone modules measure 30 cm by 60 cm, a submultiple of the framework of the supporting vaults, and the whole garden is arranged following this grid), which passes from white on the church side to black on Garibaldi square. The landscape is built in a gentle and random way. The regular network of planting is conditioned by the supporting structure. The planting on the church side is dense, and becomes progressively lighter on the basalt side. These two effects thus create a balance in terms of the brightness of the site.

03 城市公共空间 Urban public spaces

拱廊广场 *Arcades Square*

PÉNA & PÉÑA PAYSAGISTES

地点：法国埃佩尔奈
完工日期：2008-2012
面积：0.33 ha
业主：埃佩尔奈市政府
照片版权：Christine & Michel Péna

Location: Épernay, France
Completion date: 2008-2012
Area: 0.33 ha
Client: Épernay Town Council
Photo credits: Christine & Michel Péna

01. 植物在广场上以低调的方式出现，以便彰显喷泉的重要性
02. 灌木丛点状分布在地下停车场的上方
03. 这个点状散置的绿化系统将广场上原有的几棵椴树也包含了进去

01. The planting takes a step back in order to give centre stage to the fountains
02. Patches of shrubs above the underground car park
03. The string of lozenge-shaped beds incorporates the existing lime trees

某些城市寻找自身的特点，然而对其他城市而言，例如埃佩尔奈，其特色是自然形成的。游客漫步来到拱廊广场，会发现一个特殊的城市景观，此空间既古典又充满当代感。新的整治方案必须与这个场所精神相结合：明亮的石材以及带有点点花絮的绿丛，犹如一个个色彩缤纷的植物岛屿点缀着原本严峻的偌大空间。

犹如香槟酒可以同时是精确严密而充满节庆欢愉的、别致而易于亲近的，这个广场的整治也同样期望表达出如此的双重性格。微微闪耀的石材基调严谨地勾勒出整体空间的几何框架，其明亮的色调与周围石灰岩拱廊达成和谐，并且柔化了建筑上层立面的砖红，形成色彩上的互补。广场边缘近乎纯白的石灰岩，加强了其广大长方形的轮廓，突显出这个完全步行的空间。一系列种植橘树的箱盆再度强化出边缘地带的散步区，人们在此漫步，得以探索这个既是广场又是花园的新景观。景观探索在一丛丛花团锦簇的小灌木群之间展开，自然延伸到广场尽头随机排列的几棵原有椴树之间，呈现出盎然生气。

固然物理应力影响了石板的尺寸选择，然而此变化其实也传达了地面的不同使用性，造成石材规格的微妙转变。一些形体柔和的植物岛屿分布在整个广场上，犹如散落的绿色水滴，其不规则的外形与布置让人们可以自由地往各个方向行进。漫步在这些犹如彩色大卵石的植物丛之间，还可以看到一些高出绿丛的水柱喷泉在阳光下闪耀着，并听到细细水声。

Some towns are searching for their identity, while for others, like Épernay, this comes naturally. An urban landscape linking classicism and contemporary sensibilities opens up to the walker who arrives at Arcades Square. The new treatment therefore had to continue in this spirit: luminous stone above all, floral masses offering patches of colour and playing with the rigid space like plant islands on a large canvas.

Champagne has several facets: it can be both strict and festive, demanding and fun, chic and convivial, and the treatment of the square tries to express this. The framework of slightly mirroring stones draws a strict geometry. Its light colour allows one to play with the limestone arcades and soften the red colour of the bricks through chromatic complementarity. A border of almost white limestone redraws this huge rectangle, thus marking out the space entirely given over to pedestrians. Orangery planters further highlight this peripheral walk – a stroll overlooking this new landscape that is both square and garden. It evolves among the masses of flowering shrubs, and the motif naturally extends the random arrangement of the existing lime trees, in a space where the life of things is not forgotten.

Obviously, mechanical requirements conditioned the dimensions of the paving stones, which vary according to the use of the ground. The format of the stones undergoes a subtle transformation. Islands with soft forms have taken over the whole square. These plant droplets are dispersed and their apparent disorder allows for free circulation in all directions. Walking among these great coloured pebbles, you hear the sound of water, which comes from the lines of light playing above the stone masses.

04. 点状灌木丛犹如散落在广场中的糖片
05. 在椴树下乘凉
06. 穿越广场的街道以较小的石块铺面来呈现差异
07. 地面处理的轻微转变即可创造出这些水柱喷泉

04. The lozenge-shaped beds are strung like beads throughout square
05. Under the lime trees
06. The crossing is marked out by cut stone modules
07. Slight modulations in the ground allow these fountains to flow without flooding the square

03 城市公共空间 Urban public spaces

Perret Square
贝瑞广场
LAURE QUONIAM

地点：法国亚眠
完工日期：2008
面积：15 000 m²
业主：亚眠大都会联合组织
照片版权：Laure Quoniam

Location: Amiens, France
Completion date: 2008
Area: 15 000 m²
Client: Amiens Métropole
Photo credits: Laure Quoniam

01. 广场夜间景观
02. 一片大型而中空的玻璃屋顶笼罩着贝瑞广场
03. 上层广场：波浪状黄杨丛的近景

01. The square at night
02. Glass roof covering Perret Square
03. The upper square: close-up view of the waves of box

带着著名的高塔地标、面积15000平方米的贝瑞广场是亚眠城市的历史性空间，也是与火车站产生紧密关系的场所，其周边的建筑物为建筑师奥古斯特·贝瑞在第二次世界大战之后所设计建造。然而此广场却逐渐成为停车场以及火车站接送客人的临时停车点，于是市政府希望将它改造成为能够符合大城市形象的公共广场，并且将车站上下两层空间与广场连接起来。

广场上层设置了两条通往车站"失足庭"的坡道，它们与广场两侧组织机动车与公车流线的街道相连接。为了能够使人们直接抵达月台，广场的下层以缓坡的形式从上述两条坡道的高度慢慢往下延展，仿佛一个大型地毯，中央点缀了一组水柱喷泉。

一个高大的玻璃屋顶结构围绕着广场四周，但与周边建筑保持一定的距离。这个高耸的玻璃结构仿佛一个经过像素化的林下植物群，其中央的开口使得地面的音乐水柱喷泉区得以享有自然天光。一些裁剪成波浪状的黄杨丛在大玻璃罩下伸展，为广场带来生气。广场里还设置了特地为其设计的座椅、垃圾桶、照明设备等一系列的城市小品。

With its tower as a focal point, the 15,000 m² Perret Square is a historic space in constant use by people arriving at and leaving the station. The buildings that frame it were built after the Second World War by the architect Auguste Perret. As it was mainly used as a car park and drop-off point for the station, the city authorities wanted to convert it back into a public space worthy of a city. A new approach to the layout links the two levels of the station with the square.

On the raised level, the upper square is composed of two access ramps that lead to the "hall of forgotten steps". They connect to the lateral streets car and bus lanes. On the sunken level, offering direct access to the platforms, the lower square rolls out in a gentle slope between these two ramps like a wide carpet, enlivened by a central dry fountain.

A glass roof covers the whole square, without touching the facades of the buildings. The glass panels represent a pixelated image of an underwood. An opening in the centre allows overhead light to flood in over the musical fountain. Box hedges clipped into wave-like forms have been planted under the glass roof. A range of urban furniture, benches, bins and lighting, has been specially created for the square.

04. 下层广场的水柱喷泉
05. 面对着贝瑞塔楼的喷泉
06. 上层广场的波浪状黄杨丛
07. 广场夜景
08. 广场剖面图

04. The fountains of the lower square
05. The fountains in front of Perret Tower
06. The waves of box on the upper square
07. The square at night
08. Section of the square

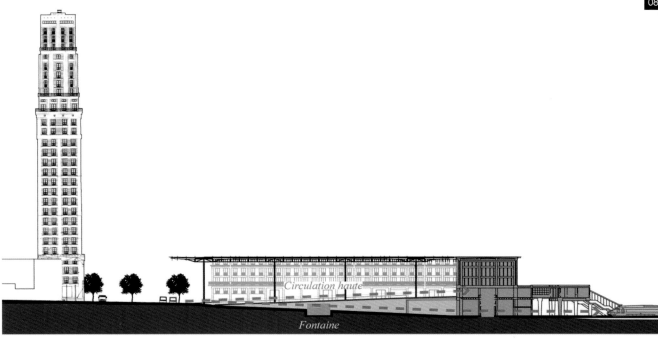

03 城市公共空间 Urban public spaces

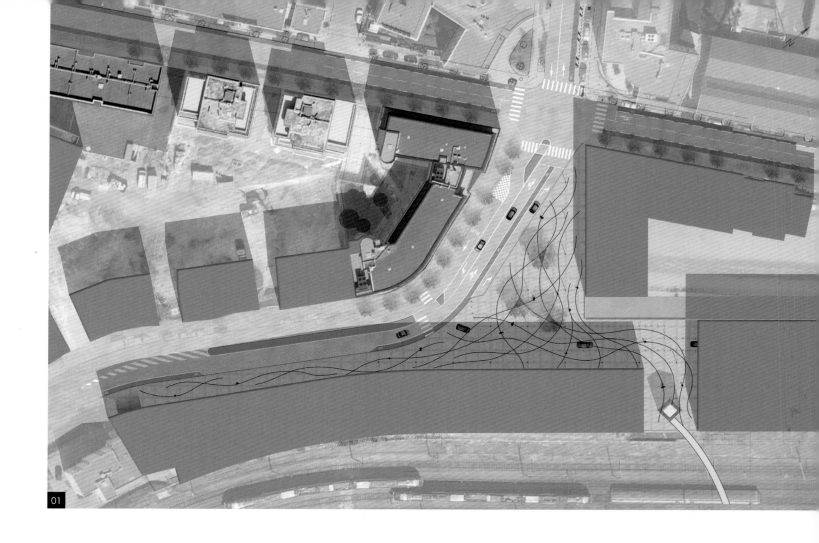

兰斯火车站站前广场
Reims Station Forecourt
SAVART

地点：法国兰斯
完工日期：2010
面积：3 465 m²
业主：兰斯市政府
照片版权：Marc Soucat

Location: Reims, France
Completion date: 2010
Area: 3 465 m²
Client: Reims City Council
Photo credits: Marc Soucat

01. 整体平面配置图：随意弯转的路径
02. 在广场所有的使用性上头叠加了曲线状的漫步路径
03. 以规律的铺地作为共通的基准坐标

01. Master plan, an organisation of random paths
02. The curved walk superimposed over all the square's uses
03. Regular paving as common ground

兰斯火车站站前广场整治方案的设计灵感来自于多位艺术家对"城市移动"这个主题所进行的创作。这些艺术家将步行流线的不规则性凸显出来，也就是说展现了弯曲的路径和规则的框架之间令人讶异的强烈对比，在城市这样一个经过理性控制的环境里，照理所有可能的流动路径都已经被纳入构思当中。

这个灵感使设计方案将两种元素结合在一起：传统规划的产物以规则的网格形式展现（让使用者感到安心）；来自当代艺术家的一些现实形态以交错的黑色弧线展现（让人们产生质疑和情绪）。它强化了行人移动路线的不确定性，他们的怀疑、方向的改变、他们的心情以及每个独立个体的自主选择。

因此，路径互相交错、回避、重叠，表现了路人的犹豫。在曲线相交处，出现了三棵树，它们从地面伸展出来，或许就是它们将地面撕裂，画出了这些深色的曲线。

The project for Reims station forecourt is inspired by the work of several artists on the theme of urban mobility. These artists show the unpredictability of pedestrian routes, the astonishing contrast between sinuous routes within a well-ordered framework such as the city, where you think you must have imagined every possible route.

These principles led the landscape architects to combine two elements: the traditional layout (the sight of which reassures visitors) represented by a regular grid, and today's new forms that come from contemporary art (which questions and creates emotions), represented by an interlacing of black curves. The resulting patterns show the uncertainties of pedestrian mobility, its questioning, its changes of direction, its moods and the choices that each individual makes.

So the paths cross each other, avoid each other, and are superimposed in a representation of pedestrian hesitations. Where the curves cross three trees emerge, rising up from the ground as if they have cracked it open to ground to create these dark curves.

04. 种植在曲线相交处的皂荚树
05. 广场上的连续动态
06. 树下的座椅
07. 从高处俯视所见的图案
08. 剖面图：黑白线条的不规律分布

04. Honey locust trees at the crossroads of the curved lines
05. Continuous movement on the square
06. A seat beneath a favourite tree
07. A design to be viewed from above
08. Section: an irregular distribution of black and white lines

03 城市公共空间 Urban public spaces

贝里运河 *Berry Canal*
TN PLUS

地点：法国谢尔省
日期：设计竞赛首奖2009
面积：190 km
业主：谢尔省议会
合作设计师：Stéphane Lemoine / AP5
图片版权：TN Plus

Location: Departement of Cher, France
Completion date: competition winer 2009
Area: 190 km in length
Client: Cher Departmental Council
Co-project manager: Stéphane Lemoine / AP5
Image credits: TN Plus

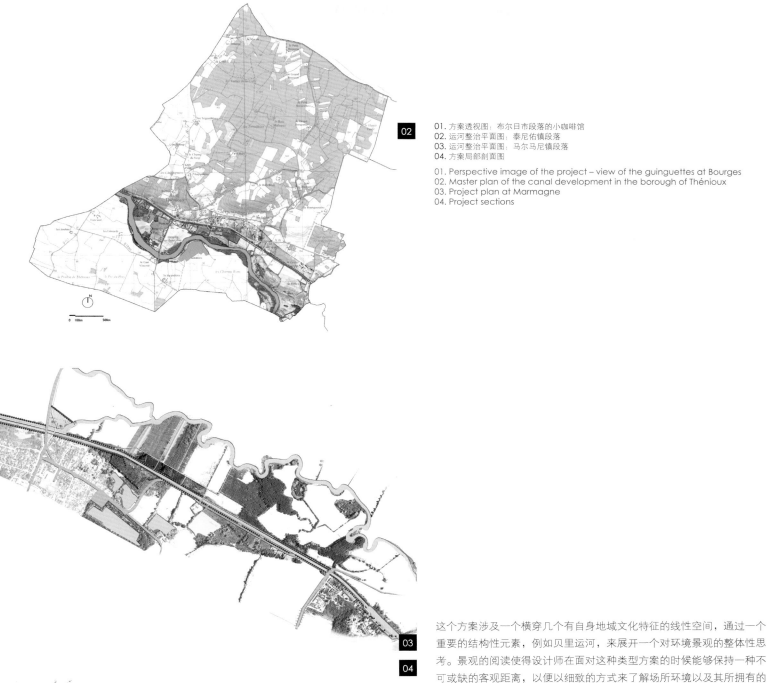

01. 方案透视图：布尔日市段落的小咖啡馆
02. 运河整治平面图：泰尼佑镇段落
03. 运河整治平面图：马尔马尼镇段落
04. 方案局部剖面图

01. Perspective image of the project – view of the guinguettes at Bourges
02. Master plan of the canal development in the borough of Thénioux
03. Project plan at Marmagne
04. Project sections

这个方案涉及一个横穿几个有自身地域文化特征的线性空间，通过一个重要的结构性元素，例如贝里运河，来展开一个对环境景观的整体性思考。景观的阅读使得设计师在面对这种类型方案的时候能够保持一种不可或缺的客观距离，以便以细致的方式来了解场所环境以及其所拥有的生态活力，进而将其明晰化，使公众能够识别出它们的特性。

这条充满生命力的崭新动脉摆脱了机动车道路的干扰，为各个村庄建立了直接的联系。通过这个管道，地区性的交通可以朝真正舒适和安全的软性交通发展。这个190千米长的线性"公共空间"将成为一个能够容纳多种实用功能的生活空间，它将涉及交通、商业活动以及城市形象和城市文化的建设。

This project was an opportunity to bring together the laying out of a long linear landscape that crosses well defined land areas, and a more universal reflection on the reading of the landscape through a strong and structuring element like the Berry Canal. It is a reading of the landscape that allows one to take the step back that is indispensable for this type of project, and facilitates a fine approach to places and their dynamics, in order to make them explicit and therefore identifiable by the public.

This new life-giving artery will permit direct liaisons between the villages by freeing up the traffic routes. In this way local journeys will be able to develop via soft transport with a real comfort of use and security. This 190 kilometres linear "public space" is a space for living where multiple practices are to be put to work. The interventions proposed address transport and commercial activity but also images and culture.

05、06. 方案透视图：艾奈镇段落
07. 方案透视图：默恩镇段落
08. 方案夜间透视图：欧吉镇段落的艺术性景观
09. 方案夜间透视图：谢尔河畔夏朗东镇段落的艺术性景观

05-06. Perspective image of the canal at Ainay
07. Perspective image of the canal at Mehun
08. Perspective image of the art installation at Augy, at night
09. Perspective image of the art installation at Charenton-sur-Cher, at night

贝里运河的景观设计必须发挥极大的想象力，显示出各种不同的尺度和视野。这个方案必须建立在一个有个性的地理环境当中，只有在一个整体的景观环境里它才能自行生存、被理解和接纳。TN Plus事务所对运河进行了整体的构思，使每个河段之间都产生相互回应。地形边界是大自然永恒持续以及它融入大地的印记，必须要注意的是，景观在任何时候都不会呈现完全的线性状态：运河也不会衍生出一个与其完全平行的景观。运河的景观是在不停发展、变化的，借此向人们展示生物与环境的生长和变化。绿化和路径的逐步建立成为了生态教学的载体。水的移动和变化、它的特质、它对植物群和动物群的影响，也都是这个景观研究和规划的重点内容。

The landscapes of the Berry Canal inspire the imagination and reveal the diversity of scales, of view points. Here, the landscape is already composed, everything is there. This project must be developed on a strong geography that can only be understood and be lived in a universal landscape framework. TN Plus thinks in terms of the totality of the linear landscape and each section echoes the others. The topographical limits are the mark of the perpetuity of nature and of its inscription in the land area. It is important to note that at no point is the landscape strictly linear: the canal does not induce a landscape that is parallel to it. The landscapes of the canal are in evolution, in transformation, propitious to a teaching of evolution. The progressive constructions of its vegetation, of its paths, become the props for an educational ecology. Water, its movements, its transformations, its properties, its impact on the flora and fauna, are the parameters developed in the work.

03 城市公共空间 Urban public spaces

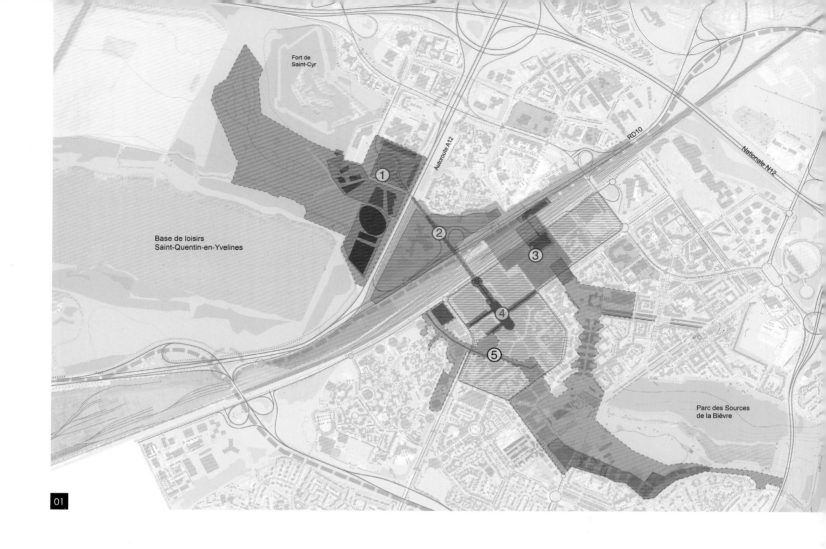

Saint-Quentin-en-Yvelines Central Urban Cluster
圣康坦-伊夫林城市中心区
TN PLUS

地点：法国圣康坦-伊夫林
完工日期：2012-2013
面积：80 ha
业主：圣康坦-伊夫林城乡区域联合组织
合作设计师：Agence Bruno Fortier
图片版权：TN Plus

Location: Saint-Quentin-en-Yvelines, France
Completion date: 2012-2013
Area: 80 ha
Client: Communauté d'agglomération de Saint-Quentin-en-Yvelines
Co-project manager: Agence Bruno Fortier
Image credits: TN Plus

01. 在前导计划当中所预定的大项目区段包括：(1) 自行车赛车场和通往休闲游乐场的空间 (2) 国道10号和其周边空间 (3) 火车站街区 (4) 乔治·蓬皮杜广场 (5) 城市运河
02. 火车站街区的5年中期计划平面图
03. 火车站街区的20年长期计划等角透视图，在车站周边建立开阔的过渡空间
04. 火车站街区改造初期入口透视图

01. Master plan showing the main sectors of the project: (1) The Velodrome and access to the leisure park, (2) The RN10 road and its surroundings, (3) The Station Cluster, (4) Georges Pompidou Square, (5) The urban canal
02. The station cluster in its intermediate phase N+5
03. The station cluster on completion, N+20: articulation via the non-built space
04. The station cluster in its initial transformation phase

圣康坦－伊夫林城乡区域联合组织属于巴黎-萨克雷国家级利益计划的策略性大区的一部分，它必须面对两个城市发展课题：如何借助此国家级利益计划所带动的大学教育界的活力来提升区域的吸引力？如何为一个带有浓厚"新市镇"（法国1960-1970年代在大城市郊区开发的卫星城市）氛围、围绕着其中央车站发展的城市重新创造出一个绿色城乡区域的特色？

TN Plus景观事务所为圣康坦－伊夫林城乡区域同时进行几个预定项目的研究，利用虽然目前被分割零散却具有潜力的土地资源，来强化这个城市化区域的吸引力。首先面对的是火车站周边区段，这是一个有如胸肺、能够带动活力的区段，需要一个彻底的重整计划来更新其特征形象。此项目主要涉及人车流动的组织、拥挤地下空间的重整和地面开阔空间的塑造，是一个目前正在实践当中的复杂项目。

运河周边是另一个决策性区段，具有强化城市活力和生物多样性的潜力。方案着重于运河河岸与河床的研究和开发，使其从连续性水池的逻辑转化成为灌溉城市的水流的逻辑。这个目前仍在研究阶段的项目的主要目标在于：为城乡区域和其地理环境建立起新的关系。

The conurbation of Saint-Quentin-en-Yvelines is situated within the large strategic sector of Paris-Saclay, designated an OIN (Operation of National Importance). This gave rise to a double problematic: how to work with the dynamic of the university facilities that had grown up on the plateau in order to develop the town's attractiveness; and how to create a new, greener identity for the sector extending from the central station, which retains the atmosphere of a "new town".

After first identifying the different project sectors, TN Plus considered several strategies to strengthen the attractiveness of this urban cluster, capitalising on the context of a land area that is high-growth even though it is particularly fragmented. First of all the station surroundings, considered as the active lung of the study perimeter, deserved a significant restructuring to truly create a new image. This area of constant movement, of congested underground areas and exposed surfaces, gave rise to a complex project that is currently underway.

The borders of the canal are another strategic sector with the capacity to strengthen urban life and biodiversity. The proposal consists of working the banks and the bed of this canal in order to move from an approach of successive pools to a water course irrigating the town centre. This proposal, still at the study stage, aims to establish a new relationship between the urban area and its geographical setting.

05. 火车站街区石砖铺面的初期研究
06. 火车站街区初期研究的三维模拟效果图
07. 火车站街区地面研究，铺地石砖尺寸的变化
08. 火车站街区石砖铺地研究三维模拟效果图，舒瓦泽勒广场
09. 火车站街区石砖铺地研究三维模拟效果图，勒德乐广场

05. Initial research on the stone plaza
06. Initial research: 3D transposition
07. Research for the ground cladding, variation in the size of the stone modules
08. 3D transposition of the different types of natural stone cladding, Choiseul Square
09. 3D transposition of the different types of natural stone cladding, Le Theule Square

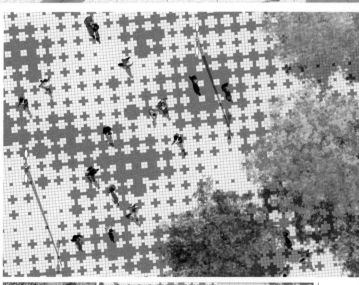
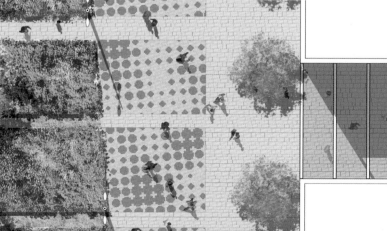

03 城市公共空间 Urban public spaces

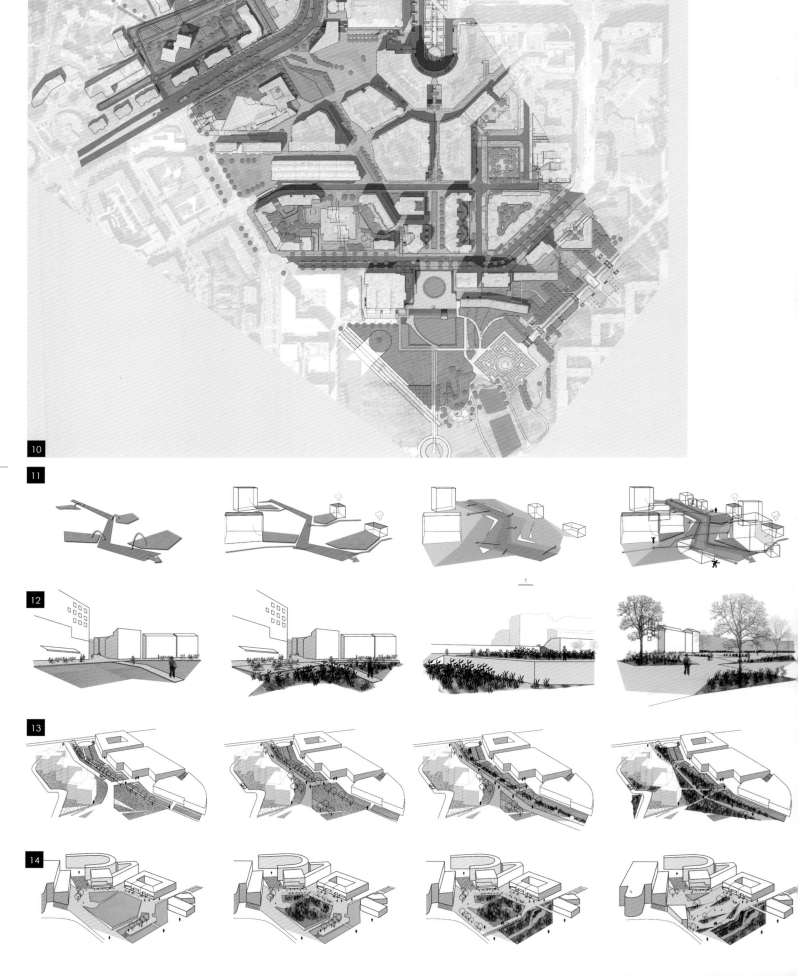

10. 运河周边研究平面图：以河岸的变化作为主题
11. 运河周边研究方案的不同阶段：延展汇水盆地
12. 运河周边研究：以穿越作为主题
13. 运河周边研究：特吕弗广场的现状和三个发展场景
14. 运河周边研究：为城市与自然创造一个新的互动关系
15-17. 运河周边研究：重新激活运河的生态环境和改造河岸空间氛围

10. Variation on the theme of the bank
11. The different time schemes of the project: extending the overflow basins
12. Variations on the theme of crossings
13. The existing state of Truffaut Square according to three scenarios
14. For a new interaction between town and nature
15-17. Biological reactivation of the canal and changes to the ambiances

03 城市公共空间 Urban public spaces

01

解 放 广 场
Liberation Square

TN PLUS

地点：法国特鲁瓦
完工日期：2008
面积：10 000 m²
业主：特鲁瓦市政府
照片版权：TN Plus

Location: Troyes, France
Completion date: 2008
Area: 10 000 m²
Client: Troyes Town Council
Crédits photo: TN Plus

01. 整体平面配置图
02. 流经大教堂前的水渠
03. 绿化园圃一景

01. Master plan
02. View of the canal looking towards the cathedral
03. View of the green rooms

特鲁瓦市解放广场的中央地下停车场的设置是广场周围场所获得重新整治的契机。借助停车场的施工和需要砍伐原有树木的机会，方案致力于将一个朝向大教堂的主要透视轴线显现出来，借此将设计构思引向扩大视觉范围，以连接邻近的三个花园和运河。

在这个步行流量十分可观的十字路口上，特鲁瓦市政府希望为市民建立一个亲切的公共空间，此目标成为整治方案的重要考量之一。作为市中心的入口，此广场也是支配街区之间相互连通关系的枢纽地带。水，是特鲁瓦经济发展的历史因素，方案同时也提供了一个让市民和水重新接触的机会：广场空间围绕着一个80米长的线性水流而组织，无论是白天还是夜晚，不同段落的水景为广场带来趣味而活跃的氛围。

广场上设有两种不同的植被类型：一种通过植物本身的形状和种植方式形成了带有中世纪色彩的花园；另一种则以枝叶茂盛的植物装点成为水生花园。山毛榉小树林在东侧紧紧围塑着广场，将人们的视线引向大教堂的顶部。夜晚，灯火通明的广场呈现出温暖的色调，并与围绕广场的中世纪建筑立面相互照应。

The introduction of an underground carpark in the centre of Liberation Square in Troyes, was the force behind the requalification of the area. During the construction work and the felling of existing trees, a major perspective on the cathedral opened up, and prompted thoughts of taking a larger visual perimeter into account, linking three gardens and the canal.

This development takes account of the city authorities' wish to give back to the people of Troyes a convivial public space, at the crossroads of a busy intersection of pedestrian routes. A gateway to the town centre, this square articulates the links between quarters. The project also offers the opportunity to rebuild contact with the water, a historic element in the economic development of Troyes. The square has been organised around a long thread of water (80 metres), fun and enlivened by changing lighting sequences both by day and night.

Two plant typologies stand out: a garden with medieval resonances in its form and its plantings, and a water garden with lush foliage. Beech groves hem in the square in its eastern part, and direct the gaze towards the cathedral tower. At night the almost extravagant lighting asserts its warm colours, answering the strong presence of the medieval facades that surround the square.

04. 流向埃米尔·左拉路的水渠景观
05. 流向埃米尔·左拉路的水渠夜景
06. 围绕着水渠的表演
07. 穿越水渠的过桥细部

04. View of the canal looking towards rue Émile Zola
05. Night view of the canal looking towards rue Émile Zola
06. A performance around the canal
07. Surface detail of the footbridge crossing the canal

371

03 城市公共空间 Urban public spaces

共和广场

Place de la République

TRÉVELO & VIGER-KOHLER

地点：法国巴黎
完工日期：2013
面积：2 ha
业主：巴黎市政府
图片版权：TVK (n°06-08), TVK + My lucky pixel (n°01, 09-11), TVK + My lucky pixel + AIK Yann Kersalé (n°5), TVK + Martin Étienne (n°02-04)

Location: Paris, France
Completion date: 2013
Area: 2 ha
Client: Paris City Council
Image credits: TVK (n°06-08), TVK + My lucky pixel (n°01, 09-11), TVK + My lucky pixel + AIK Yann Kersalé (n°5), TVK + Martin Étienne (n°02-04)

01. 在象征共和国的雕像脚下，这个广场被回归给行人
02. 此广场是大型游行与民众结集的象征性场所
03. 节庆与文化活动
04. 新的空间使用可以延伸到夜晚
05. 夜间照明设计

01. At the foot of the statue of the Republic an esplanade has been given back to pedestrians
02. The square is a symbolic place for large popular gatherings
03. Festive and cultural events
04. The square's new uses can carry on into the night
05. Nighttime lighting

由于共和广场的特殊尺度和象征意义，它在大巴黎计划中占有非常特殊的地位。广场的整治以一个有力的决定作为基础：大幅度地扩展步行空间以直接连接广场东北边的建筑立面，因而形成了一个占地2公顷、对四周林荫大道开放的广阔公共空间。

与建筑前庭相连的两个多功能平台设置在广场的一侧，它们既与交通空间隔离也受到树木的庇护。这两个便于人们休憩和交流的平台，其三个边界都由简单的台阶围绕而成。其中一个平台上建有一个170平方米的玻璃亭，建筑的通透性保证了广场视线的畅通。

新广场的统一性来自于空间的整体构成方式：采用统一的硬质铺面和同一面向的斜坡。预制混凝土这个材料的使用则体现了方案的经济性、可持续性和现代性。共和广场是一个供大众使用的广阔空间，随时准备迎接各种大小型活动，不论是已经存在的或者有待创造的。

The exceptional size and symbolic character of Place de la République gives it a special place in the Grand Paris urban planning project. Its redevelopment is based on a strong decision: to create a pedestrian public space as large as possible directly connected to the north-east facades of the square. This vast free space opening onto the Grands Boulevards will cover an area of 2 hectares.

At the back of the esplanade, sheltered from the traffic and protected by trees, two multipurpose platforms form a continuation of the forecourt. Conducive to sitting and meeting people, they are marked by simple steps on three of their sides. One of the platforms hosts a 170 m² pavilion that is entirely glazed in order to retain a continuous readability of the square.

The unity of the future square comes from its general spatial composition: a single hard surface and a main sloping movement. The material used, prefabricated concrete, also shows an economical, lasting and contemporary approach. Republic Square is a square for the people, and is here seen as an available super-surface that can adapt itself to events both large and small, from those that already form part of the Parisian calendar to others still to be invented.

06. 具有各项功能的水景：气候、社会、游戏和美观
07. 具有历史性、象征性和人民意义的共和国广场
08. 具有高质量的交通转换站，可以是过渡场所和目的地场所
09. 一个清楚界定的公共空间，以保证各种不同使用者的舒适性
10. 种植了大量树木的广场提供人们休息空间和新用途场所
11. 人人可接近的共和国雕像（整治前被机动车道路环绕）重新找回了其真实意义

06. Water and all its benefits: climatic, social, recreational and aesthetic
07. The Republic monument is historic, symbolic and for the people
08. Qualitative intermodality: through its different levels the square has strengthened its role both as a place of transit and as a destination
09. A public space clearly redefined to ensure the comfort of everyone
10. Relaxation and new uses on a square that rejoices in many trees
11. Entirely accessible, the statue of the Republic rediscovers its identity

03 城市公共空间 Urban public spaces

奥斯特利兹堤岸
Austerlitz Quays

URBICUS

地点：法国巴黎
完工日期：2012
面积：20 000 m²
业主：巴黎港口管理局
照片版权：Charles Delcourt

Location: Paris, France
Completion date: 2012
Area: 20 000 m²
Client: Paris Ports
Photo credits: Charles Delcourt

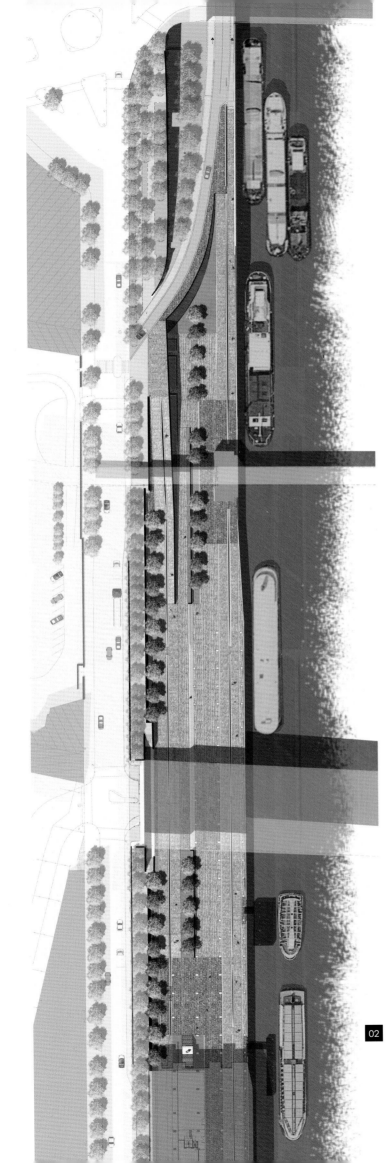

01. 此堤岸成为河左岸协商开发区的一个宽阔广场
02. 这是一个将港口活动与城市活动结合在一起的空间

01. The quay becomes the esplanade of the Rive Gauche ZAC (comprehensive development zone)
02. A space that combines port and urban activities

巴黎的塞纳河岸是被列入联合国教科文组织名录中的世界文化保护遗产，它们的整治成为城市在面对机动车的占用、决心重新收复河岸的策略之一，借此提供有益于城市活动、节庆活动与港口活动的生活空间。本方案塑造了一个以砂岩石块组成的硬质铺地空间，以与港口使用和历史尺度相协调，同时也在高密度的巴黎市中心强化出塞纳河岸的自然性格。

局部河岸的铺地石块间隙中种植了绿草，形成"缝隙"花园，此外一些树木也被重新植入基地，而爬藤植物则沿着墙面攀援而上。在巴黎港口的整治规章要求下，方案将一套符合可持续发展的设计步骤付诸执行：极力促进河流的生物多样性、妥善处理工地的废弃物、积极加强与河岸居民的协商。

尽管堤岸与河边道路的高低落差很大，项目依然坚持提高人们对公共交通的使用，因此堤岸的地面整治完全符合残疾人的交通便利，使他们能够来使用依照此项目计划所设置的水上活动驿站。此方案将一个被列为保护的历史性基地所具有的挑战与限制整合起来，并且使项目计划的落实能够与一个21世纪城市港口的功能相得益彰。

These Parisian quays, which have UNESCO World Heritage status, are the subject of a reconquest strategy that aims to replace cars with urban, festive and port activities. The project consists of developing a hard surface space in sandstone paving, in keeping with port uses and its historic dimension, while enhancing the natural character of the Seine-side location in the heart of the Parisian built environment.

In certain parts the joints of the paving stones have been planted to form the lines of a garden of interstices. Trees have been replanted and climbing plants cover the walls. Within the framework of the Paris Ports charters the project takes a sustainable development approach, which encompasses maximising the biodiversity of the river, the treatment of building site waste and in-depth consultation with the local inhabitants.

To contribute to the Seine's public transport system, the quays have been made entirely accessible to disabled users of the river transport stop put in place by the project, despite the steep embankments. In this way the project synthesises the challenges and constraints of a protected historic site and the expectations of a 21st-century urban port.

03、06. 地面的细部处理使得此空间成为一个有利于生物多样化发展的"缝隙"花园
04. 整治中（即将完工）的堤岸以及面向里昂车站钟塔的景观
05. 地面的处理限定了空间的使用功能，并且引导不同的人流动向

03&06. The detail of the hard surfaces offers a garden of interstices that encourages biodiversity
04. The quay and the view of the bell tower of the Gare de Lyon
05. Work on the hard surfaces to mark out uses and guide pedestrian traffic

03 城市公共空间 Urban public spaces

07. 高处堤岸可以通往新建的时尚中心
08. 这个场所完全规划为步行区，机动车仅限于港口使用
09、10. 高度落差的巧妙处理有利于高处堤岸和河岸道路之间的联系

07. The high quay leads to the Cité de la Mode
08. It is a pedestrian area, with motorised traffic restricted to the port location
09-10. A play of different levels enables easy links between the high quay and the banks

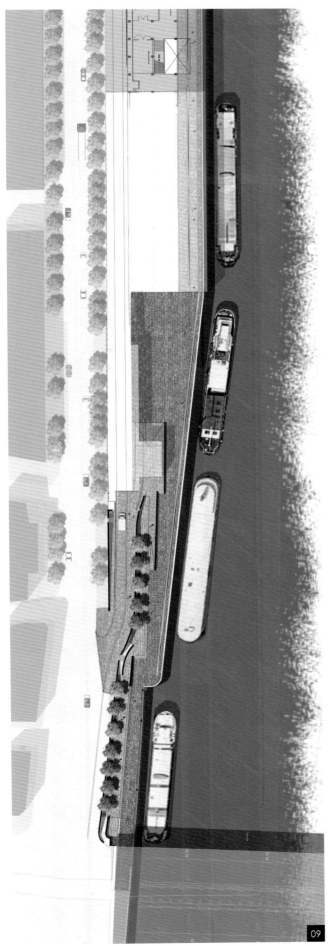

03 城市公共空间 Urban public spaces

04

Private Gardens and
Landscape Spaces

私 人 花 园
与 景 观 空 间

格雷恩城堡绿色剧场
Open-Air Theatre – Château de Grâne

AGENCE APS

地点：法国格雷恩
完工日期：2007
面积：1 000 m²
业主：格雷恩镇政府
照片版权：Agence APS

Location: Grâne, France
Completion date: 2007
Area: 1 000 m²
Client: Grâne Town Council
Photo credits: Agence APS

01. 位于旧采石场的绿色剧场
02. 以罗马式工法砌成的水泥板小径将行人迎入剧场
03. 一个对儿童同样具有吸引力的公共空间

01. The open-air theatre in the old stone quarry
02. An opus roman pathway in concrete slabs welcomes visitors
03. A public space that is also attractive for children

格雷恩镇坐落于德龙河谷的南侧,其城堡矗立于植被繁茂的山顶之上,远远就能看到它的轮廓线。在此建设一座绿色剧院显然成为势在必行的项目。场地的形势、声学质量以及镇政府希望举办文化活动以振兴市镇的愿望,这些因素促使APS事务所在对整个市镇进行全面研究衡量之后,为基地提出这个功能建议。

这座剧场看起来既简朴又优雅:通过对其风格的精准掌握和严格控制,它"自然地"融入了旧采石场的环境中,从而表现出它的形态特色。斧凿所留下的垂直岩石切面成为舞台的背景墙。剧场的入口被刻意安排在高处以便强调场景效果以及场地与作为舞台背景的山谷之间的关系。一条缓坡与基地的岩石轮廓紧密结合,不仅是带领人们进入看台的通道,同时也为观众提供了不同的观赏视角。

Seen from afar, the village of Grâne, on the southern flank of the valley of the Drôme, is distinguished by the massive silhouette of its château on the summit of a wooded hill. It was an obvious site for an open-air theatre. The configuration of the place, the acoustic qualities and the commune's desire to host cultural events in order to breathe new life into the village led the agency APS to draw up this plan as part of a larger study of how to improve the commune as a whole.

The project is both modest and elegant: through the pertinence and extreme rigour of its design it slips "naturally" into the context of the old stone quarry to reveal its morphology. The verticality of the old rock face becomes the back wall of the stage. The main entrance to the theatre has been deliberately arranged from above to accentuate the drama of the scene and its links with the valley, which forms a backdrop. A gentle ramp follows the rocky contours of the site to give access to the terraces and offers varied viewpoints.

04. 方案的"绿色"建构物融入了基地的"大自然"之中
05. 每逢夏季时节,"格雷恩星期五"为民众提供各种文化性表演节目
06. 这个"大地雕塑"设计图的精准度完全展现出了基地的地理形态
07. 向德龙河谷地理景观展开的视野

04. The "green" architecture of the project fits into the "natural" site
05. In summer, "Grâne Fridays" offer a varied cultural programme
06. The aptly chosen "earth sculpture" design reveals the morphology of the site
07. Views open up over the valley of the Drôme

方案的另一个先决条件即是通过简洁的语汇为"绿色剧场"的特色带来最直接的意义。通过材质与形体的创作,剧场看起来如同一座"大地雕塑",在一片均匀的草毯上塑造出平面和褶皱,除了确保立即的成果质量之外,更须掌控它的持久性,使其在尊重清雅结构的同时也便于维护。观众席的木质座椅板以随机的方式摆放,"飘浮"在绿色看台的中间区域,展现出大地曲线的活力。

Another motivating force for this intervention was to give true meaning to the place's identity as a "theatre in the green" through the use of a simple vocabulary. Through the physical fashioning of its materiality, the project appears as an "earth sculpture" modelling the plateaux and folds of the uniform carpet of a seating lawn. While ensuring the quality of the immediate result, this ensures its lastingness and ease of management while respecting its soft architecture. Wooden seats arranged quite randomly "float" on the central part of the green terraces, revealing the dynamic curve of the lines of the earth.

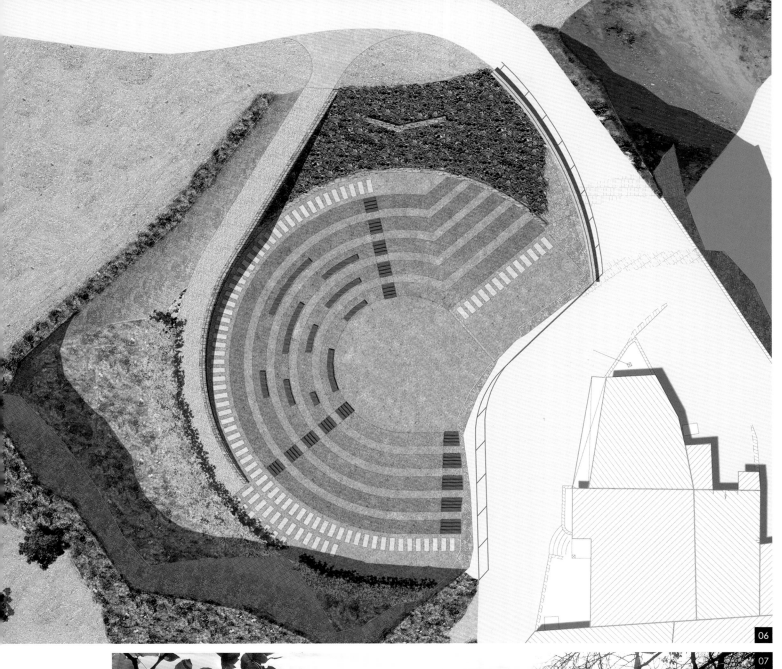

04 私人花园与景观空间 Private gardens and landscape spaces

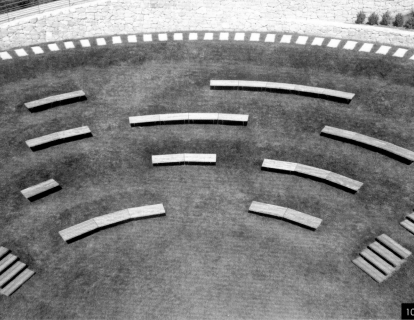

08. 从高处坡道抵达，可以欣赏绿色剧场的全景
09. 不同材质的搭配：埃佐石灰岩石墙、黄杨和禾本植物
10、11. 观众席的木质"飘浮"座板以随机的方式设置于绿色看台的中间区域
12. 方案完美地嵌入基地之中

08. Arriving from above accentuates the drama of the scene
09. A contrast of materials: walls in Eyzahut limestone, box hedging and grasses
10-11. Wooden seats "float" randomly in the middle of the terraces
12. The project fits neatly into its site

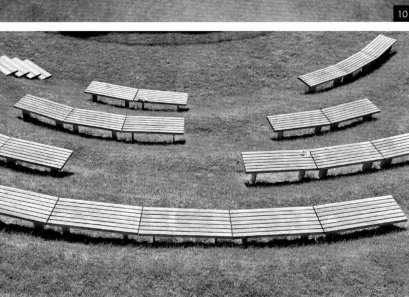

04 私人花园与景观空间 Private gardens and landscape spaces

达索系统企业园区
Dassault Systèmes Campus
ARTE CHARPENTIER

地点：法国维利兹-维拉库布莱
完工日期：2008
面积：4 ha
业主：区域地产集团
图片版权：Géraldine Bruneel

Location: Vélizy-Villacoublay, France
Completion date: 2008
Area: 4 ha
Client: Groupe Foncière des Régions
Image credits: Géraldine Bruneel

01. 整体平面配置图
02. 从办公楼的6楼平台望向"生活核心区"
03. 水渠沿岸的休憩空间
04. 轴线向西延伸，面向默东森林

01. Master plan
02. The "life heart" seen from the 5th-floor terrace
03. The relaxation space around the canal
04. In the west towards the forest of Meudon

"以另一种方式工作"是这个园区户外空间设计方案的主导理念，把室外空间塑造成露天的工作场所或者露天会议室。

南北向轴线从本方案的"生活核心区"两端延伸出去，是人们会面交流的主要空间，与镜面般的水渠平行伸展，其旁种植的树木以春季开花的树种为优先，比如李树和樱桃树。东西向轴线与毗邻的森林产生对话，花园的设计灵活而自由，间或布置着小丘，上面种植着适合在秋季观赏的树种。若干点缀着花卉的广阔草地一直延伸到基地边缘的树篱，此树篱呈现出法国西部的田园围篱的氛围。建筑间紧密联系的必要性促使设计师设置了密集的通道系统，然而为了不使园区过分充斥着硬质铺地，方案特别进行了平面图案的设计，以形成硬质铺地和绿化地面相交错的画面：虚线、折线、鹅卵石的运用…… 这些线条组合使花园的构图充满张力。

整个园区都配备了无线网络，并且提供了多种在户外工作的可能性：水渠沿岸的长凳和桌子、以锌板遮顶的凉亭、室外咖啡桌……诸多设施使人们能够方便地使用户外空间，每天在花园中无论是休息还是工作都感到舒适自在，这些都体现了一个真正园区的场所精神。

"Work differently" was the guiding line for the project for the outside spaces of this campus. The idea was to create exterior spaces as work places or seminar rooms in the open air.

The north-south axis runs through the "life heart" of the project, a central space for social activity, organised around an ornamental canal. The plantings favour spring flowering plants including sweet cherry trees and others species of prunus. The east-west axis creates a dialogue with the adjacent forest. The garden enjoys a free design, punctuated by hillocks planted with trees selected for their autumnal interest. Great stretches of flowery meadows extend as far as the hedge on the edge of the property, planted with coppice species. The necessity of connecting the buildings one to another has led to the creation of a dense network of pathways. So as to limit the number of hard surfaces, a graphic approach plays with the mineral/vegetable overlap: a play of interrupted lines, broken lines, the use of pebbles... This play of lines gives tension to the garden design.

Equipped with WiFi in its entirety, the park offers different possibilities for working outside: benches and tables along the canal, huts with zinc roofs, the tables of the outside café... This varied and plentiful furniture allows the employees to appropriate the garden, and every day reveals the constant use of these spaces for relaxing or working in the spirit of a true campus.

05. 西边轴线透视景观，朝"生活核心区"伸展
06. 公园中的工作亭之一
07. 春天花季景致
08. 穿越花园的水渠

05. The west perspective towards the "life heart"
06. One of the work kiosks in the park
07. Spring flowering
08. The canal crossing the garden

04 私人花园与景观空间 Private gardens and landscape spaces

09. "生活核心区"，硬质铺地和草坪绿毯相间搭配的地面设计
10. 人们可以在花园中坐下来开会或工作
11. 桌椅近景
12. 铺地和灌木丛
13. 不同元素形成的构图趣味
14. 硬质铺地和草坪绿毯相间搭配

09. The "life heart" and its vegetable/mineral surface
10. The employees can hold a meeting in the garden
11. Detail of the furniture
12. Paving and shrubs
13. Graphic devices
14. Mineral/Vegetable

04 私人花园与景观空间 Private gardens and landscape spaces

Garden of the IGN and Météo France
法国国家地理信息中心与气象局花园

SOPHIE BARBAUX

地点：法国圣蒙德
完工日期：2012-2014
面积：950 m²
业主：法国政府生态、可持续发展与能源部
设计总负责：Architecture Patrick Mauger
图片版权：Artefactorylab

Location: Saint-Mandé, France
Completion date: 2012-2014
Area: 950 m²
Client: Ministry of Ecology, Sustainable Development and Energy
Project manager: Architecture Patrick Mauger
Image credits: Artefactorylab

01. 种满植被的平台空间
02. 建筑物的大众入口,设有国家地理信息中心的地图商店
03. 不同的空间和用途
04. 通往餐厅的平台全景

01. Heart of the planted terrace
02. The public entrance, with the IGN shop
03. The different uses and spaces offered
04. Panorama of the terrace looking towards the restaurant

法国国家地理信息中心和气象局的两栋建筑分别由建筑师罗拉·卡尔杜希和帕提克·莫若尔建筑事务所设计,这个名为"沙丘和风"的花园犹如一条长长的飘带在两栋玻璃与木构造的建筑间伸展,并与基地的特殊地形紧密贴合。一个个种植着多年生植物、合本植物和伞形科植物的独立小花园,犹如一个个珠宝盒依次铺展开。随着季节、光线和风的变换,植被的质地、颜色与柔和外形也不断产生变化,共同编排出一系列舞蹈。三棵巨大的蛇皮槭树以其枝干的特殊质感和挺拔的身躯为花园带来别样的风情。

在这个由几个公家企业共同分享的花园中,形如沙丘的植物从为这个容纳着1600个"居民"的企业村落提供了宜人的户外公共空间,其中点缀着各种私密的或者用于分享的空间,例如一间四周围绕竹林的餐厅或者一个室外会议厅,提供人们驻足、休息、午餐甚至工作。长椅、可靠可坐的高凳、小平台、聊天角在宽阔的主通道旁形成不同的序列场景。咖啡座和建筑入口这两个居高临下的平台同样也为人们提供小憩和交流空间。

作为主要人流空间的花园以及它的使用者都倒映在周围宽大而透明的建筑立面中,真实景观与建筑立面中的倒影互相映照,并展现出天空的无穷变化,仿佛一幅永恒演变的动态画作。

A long ribbon flowing between two structures in glass and wood, one by the architect Laura Carducci and the other by the agency Patrick Mauger Architecture, the Garden of Dunes and Winds designed for the IGN (National Geographic Institute) and Météo France (the French weather service) marries with the particular topography of the site. It unfolds in a succession of immaculate units, the settings for a vegetation of perennials, grasses and umbelliferae with changing textures, colours and shapes, whose soft and flowing appearance creates a series of choreographic forms according to the variations of the seasons, the light and the wind. Three large snakebark maples punctuate the scene in all seasons with their unusual bark, giving height to the planting.

In this private park serving several public companies, architectural dunes shape the urban space of a village of 1,600 "inhabitants", punctuated by different arrangements that are either intimate or shared, where people can sit, relax, eat lunch or work, like a dining room surrounded by bamboo or an outdoor meeting room. Benches, perches, steps and conversation seats form a sequence along the wide main pathway. And overlooking this two raised terraces, one for the café and one for the entrance, offer places to take a break, chat and meet people.

Providing the main circulation space between the buildings, the garden as well as its users are fragmented in the mirrors formed by the huge transparent facades that flank it, reflecting each other and multiplying images of the sky, in successive strata, in perpetual movement and rotation.

维利兹大道企业园地
Vélizy Way Park

SOPHIE BARBAUX

地点：法国维利兹-维拉库布莱
完工日期：2013-2015
面积：9 450 m²
业主：Gécina
设计总负责：Chaix & Morel et associés
图片版权：Eddie Young

Location: Vélizy-Villacoublay, France
Completion date: 2013-2015
Area: 9 450 m²
Client: Gécina
Project manager: Chaix & Morel et associés
Image credits: Eddie Young

01. 企业餐厅的露天座
02. 整体平面配置图

01. The terrace of the company restaurant
02. Master plan

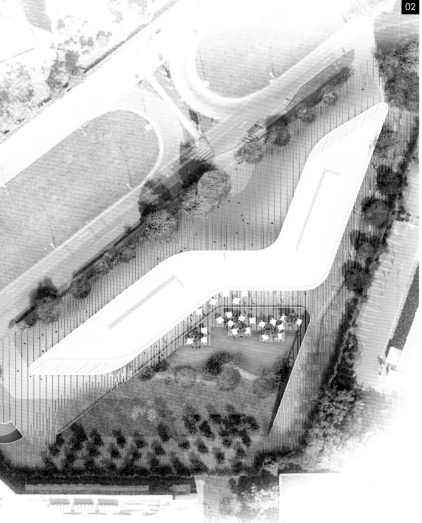

维利兹大道企业园地对于坐落于其中的现代建筑来说是一个蓝绿交织的藏宝盒，这是一栋简约、通透的新办公楼，波浪般的线条描画出流畅、动态的形体。

在基地北侧，建筑的连续转折强调出主入口和面向城市的前庭广场。此广场如同两个岛屿或两个船头之间的海峡，其低平的草皮之上点缀着高大的树木，如同点缀在莫拉纳·索尔尼尔林荫道正在施工的轻轨线路上的树木。地面的处理采用了植被和硬质铺地条状交替的形式，根据不同的通行功能时而紧密时而疏松，此线性构图引领参观者进入建筑之中……在南侧，一大片花园尽情铺展，生物多样性被列为最重要的主题。靠近建筑的一小片松林邻接着一个绿化斜沟，成为潮湿的隔离空间。它们导向种植着苹果树、梨树和樱桃树的果园，一旁则是栽培了香料的菜园，随着缓坡而向一片天然水面延伸，对面则是办公楼餐厅的室外平台。

With its water and greenery, the Vélizy Way Park offsets the contemporary, sober and transparent architecture of the new tertiary building it surrounds and reflects its undulating form and fluid and dynamic volume.

On the north side, the kink in the building marks its entrance. Here the city-side forecourt has been designed like an isthmus between two islands. The two prow-like forms seem to float above a low-level vegetation punctuated by large trees similar to those of the tramway that is being built on Maurane Saulnier Avenue. The specially designed ground treatment alternates strips of hard surface plaza and of plants. The different densities are adapted to the different types of use, creating a play of lines that invites the visitor to enter. On the south side a large garden unfolds where biodiversity is king. Here a pine grove follows the side of the building, bordered by a swale that is growing into a wetland hedge. This leads to an orchard of apple, pear and cherry trees either trained against a wall or half-standard, bordered by a kitchen garden of aromatic herbs that descends in a gentle slope towards the large natural lake in front of the terrace of the company restaurant.

03. 东西向剖面图
04. 南北向剖面图
05. 通透的建筑向户外花园敞开
06. 从莫拉纳·索尔尼尔大道抵达园区，地面的草地线条引领人们走向建筑入口
07. 企业餐厅的两个露天平台
08. 果园近景，半高树杆的果树以及一些沿着坡地边缘种植的灌木丛

03. East-west cross-section
04. North-south cross-section
05. Transparent architecture overlooking the garden
06. From Morane Saulnier Avenue, lines of grass guide visitors to the building
07. The two terraces
08. Detail of the orchard with half-standard trees and others trained against a wall along the slope

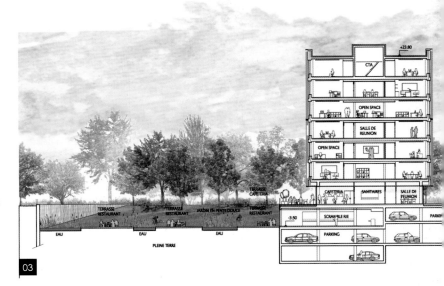

这些植物的选择是为了吸引并接纳昆虫、鸟类和小动物，并且为它们提供食物，提供一个生态栖息地，特别是成为鸟类保护协会所认可的保护站。基地的西南侧边缘种植了一长排由各种日本槭树组成的树篱，其属于灌木中较高树种，也因此呼应着基地内原本就存在的高大树种：挪威槭、欧洲赤松、西洋菩提树、欧洲黑松、樱桃李、英桐；同时也对那些新植入的树种致意：枫香树、皂荚树、毛泡桐、石松、垂柳……

The vegetation has been chosen to attract, welcome and nourish insects, birds and small animals, offering them an ecological haven. This is particularly true for birds as it aims to qualify as a Birds' Protection League habitat. A long countryside-style hedge borders the site in the south-west, sheltering a collection of Japanese maples that forms an original echo of the strata of existing mature trees on the site (Acer platanoïdes, Pinus sylvestris, Tilia x europaea, Pinus nigra, Prunus cerasifera 'nigra', Platanus x acerifolia) as well as the new ones (Liquidambar syraciflua, Glediatsia Triacanthos 'inermis', Paulownia tomentosa, Pinus pinea L., Salus babylonica pendula, etc.).

04 私人花园与景观空间 Private gardens and landscape spaces

Two Gardens of Château de Villandry

维朗德丽城堡中的两个花园

LOUIS BENECH

地点：法国维朗德丽
完工日期：2008
面积：1 800 m²
业主：维朗德丽城堡 (Henri Carvalho先生)
照片版权：Éric Sander (n°2-4, 06, 07), Georges Lévêque (n°5)

Location: Villandry, France
Completion date: 2008
Area: 1 800 m²
Client: Château de Villandry (M. Henri Carvalho)
Photo credits: Éric Sander (n°2-4, 06, 07), Georges Lévêque (n°5)

01. 花园平面
02. 太阳园：柳叶向日葵、黄盆花、蜡菊
03. 太阳园：轮叶金鸡菊、萱草、堆心菊
04. 云彩园：白花红缬草、蓝刺头、经过剪裁的永久花、玻璃菊、柳叶马鞭草和大针茅

01. Planting plan
02. The Sun garden: *Helianthus salicifolius*, *Scabiosa ochroleuca*, *Helichrysum serotinum*
03. The Sun garden: *Coreopsis verticillata*, *Hemerocallis*, *Helenium*
04. The Cloud garden: *Centranthus ruber* 'Albus', *Echinops*, clipped *Helichrysum italicum*, *Catananche*, *Verbena bonariensis* and *Stipa gigantea*

在城堡主曾祖父的方案中，这片区域内最高的一处平台被分隔成三个小花园，中间点缀着星形图案，但是这个最初的设计构思从没有被实现。这座大型梯形花园被一圈椴树所环绕，如同一块色彩斑斓的调色板铺展在苍穹之下。2006年，城堡主亨利·卡尔瓦罗委托阿历克斯·德·圣维南重新确定花园内部的主要结构线条。为了重现最初的构思，方案图中描绘了一个别具一格的星形图案，如同熠熠生辉的太阳使花园重获生机。此外，在其他的城堡内部档案文件中又发现了一个星形水池，身为历史遗迹主管建筑师的阿诺·德·圣茹昂也由此受到启发。以这张平面为基础，景观师路易·贝内什为两个花园选定了种植的花草，最终形成不同寻常的植物组合。

太阳园中，一条逐渐张开的彗星尾巴为一系列暖色禾本植物的展现拉开序幕。草茎中点缀着虾红色的或烈焰般绚丽的叶片、生长过程中颜色不断变化的玫瑰花（从最初的红色到桃色再到凋零时的印度玫瑰色）、盛开着红色或橘黄色花朵的矮灌木，还有一连串的多年生植物：从鸢尾花的火焰开始，而后燃烧着明亮的萱草、射干、山柳菊，直到严寒时节才在菊花的光彩下逐渐熄灭。

In the original project devised by the great-grandfather of the current owner, the highest terrace on the estate was to be subdivided it into three gardens arranged into star shapes. In 2006, Henri Carvallo asked Alix de Saint-Venant to revisit the main lines of the composition. To go back to his great-grandfather's sketch, the design is enlivened by a single star, a sun at the centre of the space. Other "house" archives revealed a star-shaped pool, which, in addition, inspired the architect Arnaud de Saint-Jouan (ACMH). From this plan, Louis Benech took on the task of choosing the plants for the two garden rooms employing an unusual plant combination. Today, though the original project was never achieved, this large trapezium framed by a belt of limes and hornbeams displays the colourful palette of the heavens.

For the Sun garden, a comet's tail flares in a riot of flame-coloured grasses. These stems are joined by leafy plants in prawn or flame colours; a rose that changes colour at each stage of its blooming (red, then peach, fading into Indian pink); small shrubs with red or orange flowers; and a host of perennials, igniting with the fiery irises, turning into a blaze of day lilies, leopard flowers and hawkweeds, and dying into the embers of the first frosts with the chrysanthemums.

05. 太阳园：多色大戟和圆苞大戟、金焰绣线菊、
辉煌欧亚槭、日光兰和东方罂粟
06. 云彩园：天蓝鼠尾草、珍珠菜、白天鹅松果菊、
垂枝柳叶梨、辽东楤木
07. 云彩园：倒伏荆芥、毛叶黑心蕨、裂叶罂粟、
大针茅

05. The Sun garden, *Euphorbia polychroma* and *griffithii*, *Spiraea japonica* 'Goldflame', *Acer pseudoplatanus* 'Brilliantissimum', *Asphodeline* and Oriental poppies
06. The Cloud garden, *Salvia uliginosa*, *Lysimachia clethroides*, *Echinacea purpurea* 'White Swan', *Pyrus salicifolia* 'Pendula', *Aralia elata*
07. The Cloud garden, *Nepeta mussinii*, *Dorycnium hirsutum*, *Romneya coulteri*, *Stipa gigantea*

404

云彩园内则种植了一系列从白色、银色到蓝色和紫色的植物。夏季吸收了更多阳光的苗圃将唤醒球蓟的蓝色花朵、布宜诺斯艾利斯的马鞭草、玫瑰和委陵菜的白色花朵，以及素有"清晨之光"之称的中国芒和针茅。在城堡主、一位建筑师和两位景观师的热情参与之下，这些新花园得以诞生，但若非维朗德丽城堡的园丁提供了他们卓越的专业技术，以精巧和热情的方式来照料这片花园，这些植物将无法呈现出今日的丰裕璀璨。

The Cloud garden unfolds in a chromatic range of white, silver, delicate blues and a few violets. It is in summer that the patch best captures the light, waking up the blue flowers of the globe thistles, purpletop verbenas, white roses and potentillas interspersed with *Miscanthus sinensis* 'Morning Light' and *Stipa gigantea*. Born of the combined enthusiasm of the owner, an architect and two landscape architects, these new gardens would never have seen the light with such a rich assortment of plants without the exceptional expertise of the gardeners of Villandry, who ensure their upkeep with knowledge and love.

04 私人花园与景观空间 Private gardens and landscape spaces

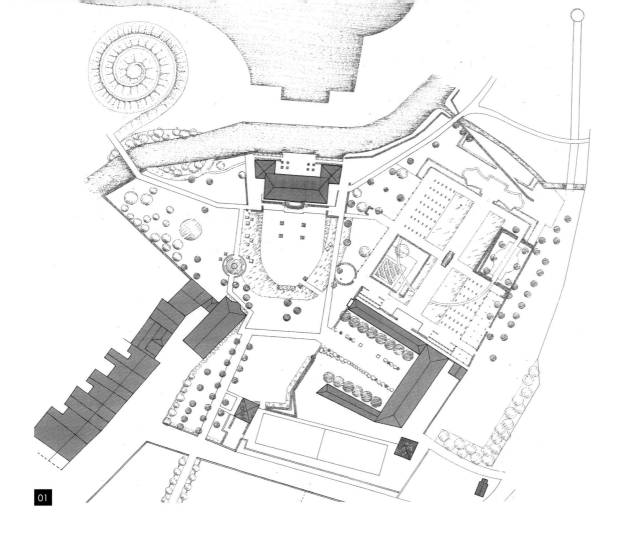

01

庞格城堡花园
Gardens of Château de Pange
LOUIS BENECH

地点：法国庞格
完工日期：2003
面积：10 000 m²
业主：庞格侯爵与侯爵夫人、摩泽尔省议会
照片版权：Éric Sander (n°2-5, 07, 08), Yann Monel (n°06)

Location: Pange, France
Completion date: 2003
Area: 10 000 m²
Client: Marquis and Marquess de Pange, General Council of Moselle
Photo credits: Éric Sander (n°2-5, 07, 08), Yann Monel (n°06)

01. 路易·贝内什花园方案平面图，2000年
02. 城壕边界种植芒草和鸢尾，前方为紫杉裁剪成的金字塔
03. 阴凉围园景观
04. 裁剪成蝴蝶状的矮篱，受到18世纪花园平面的启发

01. Master plan of Louis Benech's project, ink, 2000
02. Edge of the moat in miscanthus and irises, beyond the yew pyramids
03. View of the shade room
04. Hedge clipped in a butterfly form, inspired by the 18th-century garden plan

庞格城堡于18世纪在一座旧堡垒的遗迹上建设起来，在时间长河中它的面貌经常随着城堡主的更迭而发生变化。19世纪城堡被全面整修，二次世界大战期间先后在德军和美军的炮火轰炸之下遭受重创，后来成为收容儿童的社会机构，直到1970年代中期才成为庞格侯爵家族后代的居住之地。

路易·贝内什于2000年赢得城堡花园改造的设计竞赛，他清除了遮挡视线的障碍（停车场、足球场、一系列不受欢迎的球果植物……），使面向周围乡野景观的视野变得开敞。景观师致力于建设一座属于21世纪的花园，同时保留它与过往的关联，但不趋附于修旧如旧这种并不可靠的做法。他把考古学上的参照、实存元素、对旧平面图的解读以及未来的使用需求都联系在一起。城堡边缘的花圃造型让人们回想起马蹄形的城壕。田园风光、装饰风格、娱乐性的花园都是对18世纪城堡的回应，位于主要林荫道交叉点的十字形水池把最初设计方案中明显可见的水元素带到今日的花园中。

Built in the 18th century on the remains of an old fortress, the château of Pange has frequently changed its appearance over the years and through its different occupants: the estate was completely remodelled in the 19th century; it suffered in the hands first of the Germans then the Americans during the Second World War; it then became a children's home, before being inhabited once again by the new generation of the Marquis of Pange's descendants in the middle of the 1970s.

After winning the competition for its landscape design in 2000, Louis Benech has opened the landscape up to the surrounding countryside by removing the elements that obstructed the view (a car park, a football ground, an ill-advised planting of conifers…). He set about placing the garden in the 21st century without turning his back on the past: he thus brings together archeological references, existing elements, an interpretation of the old plans and a view to how it will be used in the future. The shape of a flower bed hugging the château evokes the old horseshoe-shaped moat. The rustic and stylised ornamental gardens echo the 18th-century plans, as does the axis of the cruciform pond, which highlights the intersection of the main paths by reintroducing water, in part inspired by the original plan.

05. 向四周乡野敞开的花园
06. 位于两条草坪通道相交处的十字形水池
07. 绿色剧场两条通道的交叉点
08. 穿越蔚蓝园圃和银色园圃的小径

05. A garden open onto the surrounding countryside
06. Cruciform pond at the intersection of two mowed avenues
07. At the intersection of two avenues of the green theatre
08. Cross avenue in the sky blue and silver enclosed garden

路易·贝内什设想了便于维护的花园形式：由禾本植物构成的方形草场以及围绕它们的草皮小径每年只需要整理两次；银色和蔚蓝色的植物构成如蔬菜园般的花园，这两种颜色也正是庞格家族徽章的颜色。这些自然而不露痕迹的改造掩盖了项目在实现过程中的复杂性，在这片历史遗迹中注入了柔美与平和的气氛，花园在建成之后即获得了法国"卓越花园"荣誉标记，在预算极其有限的条件下实现了摩泽尔省政府的这项整治措施。

Louis Benech designs patterns that are easy to maintain: the meadow squares planted with grasses and flowers and framed by strict lawn paths only have to be cut twice a year; the sky blue and silver garden – the colours of the family crest – is composed like a kitchen garden. Masking the complexity of its construction under this apparent spontaneity of gesture, it breathes softness and serenity into this historic estate, which was quickly rewarded with the label "remarkable garden". In this way it fulfilled to the ambitions of the Moselle department within the confines of a very limited total budget.

04 私人花园与景观空间 Private gardens and landscape spaces

"Eden Bio" Urban Hamlet
绿色伊甸园
MAISON ÉDOUARD FRANÇOIS

地点：法国巴黎
完工日期：2009
面积：建筑净面积 6 760 m²
业主：Paris Habitat 巴黎社会住宅机构
照片版权：David Boureau (n°9-11), Nicolas Castet (n°01-03, 05, 07, 08), Nikola Mihov (n°12)

Location: Paris, France
Completion date: 2009
Area: Net floor area 6 760 m²
Client: Paris Habitat
Photo credits: David Boureau (n°9-11), Nicolas Castet (n°01-03, 05, 07, 08), Nikola Mihov (n°12)

01. 小区内的小径
02. 户外楼梯
03. 中央建筑立面的植物帷幕处理

01. An alley
02. Exterior staircase
03. Planting on the facade of the central building

绿色伊甸园项目分布在巴黎的一个完整街坊上，其中穿越过几条几米宽的小巷。这片土地在过去曾经是巴黎城郊的菜地。建设计划包含了社会住宅、艺术家工作室、几间社团活动室和一个小餐馆。在街边加设的两个温室是为了表达对往昔菜地的纪念，住户的信箱也设置于其中，居民们得在香蕉树干和葡萄藤架之间取回他们的信件。

独立住宅叠加成公寓楼的形式，透过每栋楼各自的室外楼梯可通达每户人家。面对面的住宅立面以植物帷幕进行处理以避开视觉的侵扰，几个倾斜的室外楼梯靠在其边上。立面上的植物直接栽种在地面的土壤中，数以千计的紫藤顺着垂直的原木花架攀藤缠绕而上。在楼房的楼梯和阳台之间，每隔2米就有一个由小树枝搭成的花架，它们正在逐渐被植物所占满。整个住宅楼房展现出多种质地和形体，其样貌随着四季而更迭。

地面处理也同样受到建筑师的关注，强制将一片厚实的天然绿色植土铺在地上，并在整个施工阶段中受到保护。这些土壤如此稀罕，建筑师却决定不在其上栽种任何植物。这个决定所产生的景观是城市荒地的景观，一片未经规划的景观。它将被那些不预期而来的当地植物所占领，成为一片真正的城市荒地，自生自灭。最终的景观将充满偶然性，由那些自发地来到此地生根的野生植物所组成，借此弥补那些经过选择的植物的不足。

On a whole Parisian block, criss-crossed by narrow alleys that are a remnant of the neighbourhood's market-gardening past, Eden Bio is an "urban hamlet" encompassing a block of public housing and artists' studios, a few premises for local associations and a small restaurant. In memory of the historic fruit gardens two greenhouses have been added on the street; they serve as letter boxes, and the tenants come here to collect their post between the trunks of banana trees and a climbing vine.

The homes are made up of superimposed apartments served by their own exterior staircases. They look out on a plant curtain on which several of these staircases have been attached. The plant facade is planted in real earth. A thousand wisteria plants climb on wooden supports assembled vertically. Situated every two metres between the staircases and the balconies of the building, this scaffold of wooden stalks is gently colonised by the plants. The structure as a whole has a great variety of forms and materials that change according to the seasons.

The ground has not been forgotten. A thick layer of organic earth was laid and protected during the building works. In this rare earth it was chosen to plant hardly anything at all. The landscape that results is that of an urban wild place, a landscape that is not designed, but is colonised by unexpected indigenous plants. A real piece of nature in the city, it is self-sustaining, creating a random landscape made up of wild species that add themselves spontaneously to those that have been chosen.

04. 方案模型
05. 绿化的木结构
06. 方案模型大样
07. 独立住宅的立面被爬山虎给占领
08. 建筑脚下的荒地铺着天然绿色植土，以供植物自然生长
04. Scale model
05. Wooden structure colonised by vegetation
06. Scale model (detail)
07. Town house facades colonised by Virginia creeper
08. An urban wild place, benefiting from Demeter organic soil

04 私人花园与景观空间 Private gardens and landscape spaces

09. 整体景观
10. 由锌板、铜板和水泥组成的建筑立面
11. 夜间照明
12. 供植物攀爬的木结构夜景

09. Overall view
10. Zinc, copper and concrete facades
11. Nighttime lighting
12. Structure for plants, night view

04 私人花园与景观空间 Private gardens and landscape spaces

伯纳尔博物馆

Bonnard Museum Gardens

AGENCE HORIZONS / JÉRÔME MAZAS

地点：法国勒卡内
完工日期：2011
面积：3 100 m²
业主：勒卡内镇政府
合作设计师：Ferrero & Rossi
照片版权：Nicolas Faure & Jérôme Mazas

Location: Le Cannet, France
Completion date: 2011
Area: 3 100 m²
Client: Le Cannet Town Council
Co-project manager: Ferrero & Rossi
Photo credits: Nicolas Faure & Jérôme Mazas

01. 整体平面配置图
02. 从博物馆平台望向其前庭广场
03. 前庭广场：阴影与坐凳

01. Master plan
02. The museum forecourt from the terrace
03. The forecourt: shade and seating

伯纳尔博物馆的前身是维尔尼别墅，在仔细勘测了位于它上方的葡萄园坡地直到皮埃尔·伯纳尔别墅之后，杰侯姆·玛扎斯事务所决定保留既有的大量丘陵花园，以及其溢出到山丘小路的部分。这些花园的状态良好、易于设置，并且光线明亮，自然而然地令人想要将此氛围加以延伸，"将花园带入街道"。正是基于这个想法，杰侯姆·玛扎斯与建筑师们合作将此方案实现于城市之中，并与小山丘紧密结合在一起。

位于卡尔诺大街并拥有维尔尼花园古老棕榈树的博物馆前庭广场、水池、蔓生植物、皮埃尔·伯纳玫瑰树、杏树以及场所的宁静氛围，这些出现在花园里的元素，都反映了油画大师（皮埃尔·伯纳尔）的作品精神。景观师希望为此场所重新找到属于它自己的色彩，在此着重冷暖色调的交织，以镜面水池来强调花园的深度与博物馆西立面的退缩，此外，编织而成的围篱也成为花园的表达语汇之一。

通过这个花园和建筑的尺度，方案的构想和目标在于为公众创造一些能够呈现皮埃尔·伯纳尔作品基础氛围的空间情境。

After surveying the hillsides situated above the museum in the old Maison Vianney and running up to the house of Pierre Bonnard, the agency Horizons decided to preserve the strong presence of the hillside gardens and their overflow onto the cliff roads. The care that had been taken over the gardens, the simplicity of their design and the light that reigns here naturally led the agency to want to extend this atmosphere: "to bring the garden out into the street". Guided by this idea, Jérôme Mazas worked with the architects to work the project into the town from its starting point on the hillside…

The forecourt on Carnot Boulevard with the palm trees of the old Vianney garden, the ornamental pool/strip of water, the climbers, Pierre Bonnard's rose trees, the almond trees of course, and a feeling of serenity all had to appear here, echoing the canvases of the great painter. The agency wanted to bring back the colours of his palette, the play of warm and cold colours, and to add an ornamental lake to play with the depth and the distancing of the museum's western facade. The trellis fence also plays a part in the vocabulary of the garden.

Through this garden and the scale of the building, the idea was to give the public some of the ambiances that provided the basis for the works of Pierre Bonnard.

04. 花园的西边入口
05. 西边入口的阶梯
06. 镜面水池和坐凳
07. 镜面水池的倒影

04. The west entrance of the garden
05. Staircase of the west entrance
06. Ornamental lake and bench
07. Ornamental lake

04 私人花园与景观空间 Private gardens and landscape spaces

08. 东边的隐秘花园
09. 隐秘花园，通向博物馆平台
10. 隐秘花园，低矮植物的组合
11. 隐秘花园，在棕榈树的庇护下

08. The discreet garden to the east
09. The discreet garden, view towards the terrace
10. The discreet garden, low-level plant composition
11. The discreet garden, under the palm trees

04 私人花园与景观空间 Private gardens and landscape spaces

小姐妹教会花园
Garden of Little Sisters of the Poor
AGENCE HORIZONS / JÉRÔME MAZAS

地点：法国土伦
完工日期：2011
面积：5 500 m²
业主：小姐妹教会
照片版权：Jean-François Ravon

Location: Toulon, France
Completion date: 2011
Area: 5500 m²
Client: Les Petites Sœurs des Pauvres
Photo credits: Jean-François Ravon

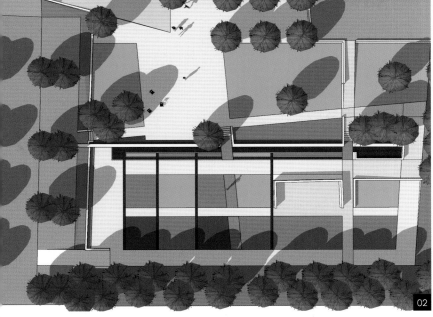

01. 水池全景
02. 整体平面配置图
03. 水池与过桥
04. 老墙与水池

01. General view of the pond
02. Master plan based on a 3D model
03. The pond and the footbridge
04. The old wall and the pond

方案的基地原先是一处农业用地，拥有一座刚被翻新的小教堂和一些教会建筑，这块土地具有一种家庭农作的风格：隆长的石墙、水泥的棚架、两口水井和一个养着几条金鱼的水池。几栋集体建筑包围着地块，对它产生极大的影响。

方案保留了传统的农业特征：通过食品农作物（例如果树）的种植，以及一块为疗养院的病人所准备、作为园艺工作室的小菜圃，石墙和水泥棚架得以被保存下来；水井中的水用来浇灌一些经过挑选、需水量低的植物；棚架在两个地点被复制使用，以便产生阴影空间，并塑造出更为私密的场所。一个新水池沿着老墙而设立，经过调整之后，可以利用重力来灌溉蔬菜地。

如今，新植的植被还不够成熟，仍然需要生长才能达到初始的野生氛围，但是一个繁盛花园的基础已经被建立了起来：植物被密集栽种，鸟儿们大快朵颐，既有的树木伸展出美丽的绿荫，流水带来了一丝清凉和永久的动感。

On an old agricultural plot where a chapel and congregational buildings have been renovated, the site has a small-scale farming character: a long stone wall, an old concrete arbour, two wells and a fish pond stocked with goldfish. The plot is surrounded by blocks of public housing that have a strong visual impact.

The project helps preserve the ancestral agricultural character of the place: the stone wall and concrete arbour have been retained thanks to the rationale of planting food-producing plants such as fruit trees and a small kitchen garden to be used for a gardening workshop by the inhabitants of the retirement home. The wells are employed to water the plants, which have been chosen for their low water consumption. The idea of the arbour was repeated in the form of shady, more private areas in two places. A new pond has been created along the old wall and arranged to provide gravity-powered irrigation for the kitchen garden plots.

Today, the planting is still too recent and needs to grow in order to regain the rather wild ambiance that was here at the outset, but the roots of a flourishing garden are there: the planting is dense, birds love it, the trees that were already established provide a delightful shade and the water brings a freshness and constant movement.

05. 水渠中种植了鸢尾、勋章菊和禾本植物
06. 花园边缘一景
07. 花园剖面图显示出了昔日的农业梯田地形
08. 在梧桐树庇荫下的花园
09. 在种植椴树之前的椅凳

05. The canal planted with irises, gazanias and grasses
06. Beside the garden borders
07. The section of the garden shows the low walls of the old agricultural plots
08. Beds in the shade of the plane tree
09. Benches in front of the lime tree plantations

JARDIN D'ACCUEIL   JARDIN PÉDAGOGIQUE

04 私人花园与景观空间 Private gardens and landscape spaces

10. 水景附近的植物组合
11. 神甫花园
12. 老梧桐树下的植物
13. 水池细部

10. Plant composition near the well
11. The curate's garden
12. The shade of the old plane tree
13. Detail of the pond

04 私人花园与景观空间 Private gardens and landscape spaces

阿朗松大学校园

Alençon University Campus

L'ANTON & ASSOCIÉS

地点：法国阿朗松
完工日期：1997-2012
面积：21 ha
业主：卡昂学院教育机构、奥恩省议会
照片版权：Agence L'Anton & Associés

Location: Alençon, France
Completion date: 1997-2012
Area: 21 ha
Client: Education Authority of Caen, General Council of Orne
Photo credits: Agence L'Anton & Associés

01. 整体平面配置图
02. 新建筑边缘景观
03. 大型绿化斜沟、人行道和小广场
04. 石笼长凳,其上的坐垫则为回收塑胶材料制成

01. Master plan
02. Overall view of the surroundings of the new buildings
03. Large swale, walks and small squares
04. Gabion benches, with seats in recycled plastic

位于乡间的阿朗松大学校园是逐步建设起来的,因为它先后汇集了私立、半公立和公立的教学机构。1997年,在建设四栋新校舍的机会下,奥恩省议会委托朗东景观事务所对大学校园进行重新规划。继校园总体规划以及建立相应的城市和建筑规范之后,室外空间整治于2012年完成。

这个过程需要与使用者保持紧密的联系、进行深入的协商,他们的平常习惯已经占据了所有可使用的空间。而新提出的总体规划着重基地的整体性,在校地边缘建设一条服务性道路,将那些包围着校园的建筑物联系起来。

Located in open countryside, the university campus of Alençon grew up from an aggregation of private, para-public and public educational establishments over several years. In 1997, the Orne department commissioned a study with a view to restructuring the campus at the same time as adding four new university facilities. After drawing up a site master plan and the urban and architectural prescriptions, the laying out of the exterior spaces was completed in 2012.

This process required a close consultation with the users, whose habits had strongly colonised the available space. The proposed masterplan organises the whole site from the starting point of a peripheral access road that encloses the buildings, which themselves protect the campus.

05. 一栋教学楼前的入口小广场
06. 作为交流空间的小广场
07. 坐凳点缀着空间也衬托出其使用性

05. One of the small squares at the entrance to the establishment
06. The small squares are spaces for social interaction
07. Benches punctuate the space

这个根据HQE高环境质量标准而完成的任务,达到了以下目标:
- 重新赋予校园整体性。为短期和中期建设的设施提供最优化的配置:大学餐厅、图书馆、行政楼、第一阶段教学楼和教师培训机构教学楼
- 促进学生之间的交流与融合。改善公共空间的质量,并建设一个名副其实的校园中心
- 建立一个高效率的交通和停车系统,便利人们抵达校园和各个校舍。为不同交通行进方式建立具有互补性的系统和清楚的空间等级:行人、自行车、机动车和大型货运车
- 提供一个景观与建筑上的强烈识别性。不仅使校园本身具有特殊意象,同时也成为阿朗松城乡区域西边入口的象征

This mission, which adheres to High Environmental Quality (HQE) standards, enabled us to offer:
- A greater coherence for the site, optimising the installation of short and medium term programmes: a university restaurant, site library, administration, general studies degree faculty, teacher training institute, etc.
- A setting that makes it easier for students to intermingle and meet, through improving the quality of the social spaces and giving the site a real heart,
- A system of circulation and parking that is functional and efficient in order to make the access to the different establishments and to the campus itself more legible. This implies a complementarity and a clear hierarchisation of the different forms of transport: pedestrians, cycles, automobiles and heavy goods vehicles.
- A strong landscape and architectural identity, on the scale of the site and on the scale of the western approach to the Alençon urban area.

04 私人花园与景观空间 Private gardens and landscape spaces

01

法国大使馆花园

Gardens of the French Embassy

FLORENCE MERCIER PAYSAGISTE

地点：中国北京
完工日期：2011
面积：1 ha
业主：法国外交部
项目代理建筑师：SAREA Alain Sarfati Architecture
照片版权：Florence Mercier

Location: Beijing, China
Completion date: 2011
Area: 1 ha
Client: French Republic – Ministry of Foreign Affairs
Project represevtive: SAREA Alain Sarfati Architecture
Photo credits: Florence Mercier

01. 枫树林荫道成为中央大草坪的框架
02. 广大的草坪可用来作为举行接待会的空间
03. 禾本植物的层叠与略为下沉的小径

01. Avenues of maples frame the central lawn
02. The main lawn, a space for holding receptions
03. Systems of folds and sunken paths

位于北京的法国大使馆花园是一个生活的场所，也是供人观赏和表达法国文化的场所。通过与东方文化相融合的场所布景和具有代表性的法国景观片段，方案展示了一系列具有对比效果的植物环境。

随着步行者的节奏，一个空间逐渐转变为另一个空间，轮流成为舞台场景和舞台背景。中心花园的绿毯和成排的枫树让人想起巴黎的皇家花园，并成为一个举行接待会的空间。禾本植物的层叠为大草坪带来生气，并形成一种色彩的颤动，反射着天空的光线，仿佛长满了鲜花的田野。

在场景的端头，灌木花园由一个围绕着三块空地的环形散步道组成，这三块小空地提供了观赏水幕和水池的不同的视点。在这些起伏的空间中，大片的蕨类植物和高大树木的节奏形成了一种过滤效果，并使光线变化多端，让人联想到法国森林景观的画面。位于边缘的花园种植了不同层次的球果植物、彩叶树木和浆果树，形成了一些较为私密的空间，展现了边缘树林的氛围

The gardens of the new French embassy in Beijing offer places for living, for contemplation and for the expression of French culture. Through a scenography that works in collusion with Oriental culture, the project reveals a series of contrasted plant atmospheres that refer to fragments of French landscapes.

As the walker moves through them, one space slides into another, building the scene and the backdrop in turn. The central garden with its green lawn and its rows of maples recalls the gardens of the Palais-Royal in Paris and forms a space for holding receptions. Folds of grasses enliven the main lawn and create a coloured vibration that reflects the light of the sky, evoking the fields of flowers.

In the distance, the underwood garden is made up of a circular walk around three clearings which offer different viewpoints on the waterfall and ornamental lake. In this undulating space, the carpet of ferns and the rhythm of the majestic trees create filtering effects and a play of light, like French forest landscapes. The border garden, given depth by the combination of conifers, trees with coloured leaves and small berry trees, hosts more intimate spaces and evokes the atmosphere of wooded borders.

04. 林下花园中的小径
05. 从林下花园望向高处花园
06. 中央大草坪周边的林荫道
07. 大草原平边缘的禾草层叠令人联想起乡野景致
08. 层理与线条的搭配

04. Path through the underwood garden
05. The high garden seen from the underwood garden
06. Shady avenues around the central garden
07. The folds of the main lawn evoke fields
08. A play of lines and strata

04 私人花园与景观空间 Private gardens and landscape spaces

Garden of Carpe Diem Tower
卡尔普·迪尔木大厦花园
MUTABILIS

Location: La Défense, France
Completion date: 2012
Area: 1 000 m²
Client: Aviva
Image credits: Mutabilis (n°02, 03, 05, 07, 08, 10-14), L'autre image (n°01, 09), Stern&SRA (n°04), Quentin Garel (n°06)

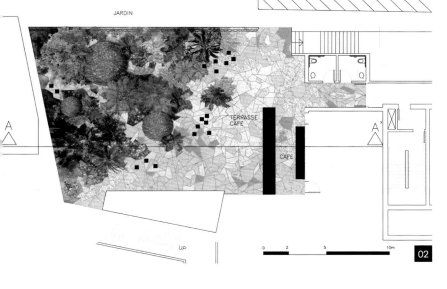

01. 神奇花园透视图，建筑大厅入口
02. 大厅平面图
03. 神奇花园剖面图
04. 卡尔普·迪尔木大厦的底层空间，成为商贸区外环道与中央楼板平台广场之间的连接点

01. Perspective image of the Garden of Marvels, entrance hall
02. Plan of the hall
03. Section of the Garden of Marvels
04. At the foot of Carpe Diem Tower, the boulevard and the plaza form a staple

卡尔普·迪尔木大厦位于拉德芳斯商务区，在这个著名街区的改造建设中，它是新一代高层建筑的象征。在高层建筑中建造花园的问题从来也没有被真正探讨过。尽管"绿色设计"有些成功的范例，但是空间和花园的概念总是不存在。

穆塔彼利斯事务所想要在此建造一座花园，而不是一个绿色的附属物，这就需要回到花园的起源。如何在四周高楼耸立的环境中建立一座围合花园的感觉？这需要将花园建造在"钟形罩"之下，同时定义一个空中花园的概念，因为在拉德芳斯这个建立在楼板平台上的商贸区里，花园原本就是脱离了实际地面的。景观师从这两个想法出发，逐步完善了设计方案。

穆塔彼利斯事务所期望为这座大厦创造一个能够为人们带来惊叹和生产愉悦感觉的花园。这个花园犹如一座想象中的热带丛林，它跨越了所有的界限，蕴含着大自然中的美妙事物：攀缘的、悬挂的、盘绕的、蔓生的…… 仅仅是它出人意料的异国情调就足够使人眼花缭乱。热带丛林的想法由对垂直组合和尺度关系等细节的研究而获得实现。

In the business district of La Défense, Carpe Diem symbolises a new generation of towers forming part of the renewal and redynamisation plan for the famous district. The question of garden in a high-rise construction is never truly tackled. At the very most there is "green" design, which is sometimes successful, but the notion of space, of garden, is always absent. Mutabilis wanted to construct a garden rather than a green accompaniment, and that meant returning to the origins of the garden.

How could they create the feeling of *hortus conclusus* in the shadow of the monumental world of the towers? Mutabilis had to put the garden under a "cloche", and in parallel to define the notion of a hanging garden because the gardeners were going to work without soil. It was from these two ideas that Mutabilis literally grew its project.

The garden was designed to bring real pleasure to the tower by creating a sense of wonder. An imaginary jungle, it pushes all the limits and contains nature's marvels, which climb, hang, wrap around or creep... Its unexpected exoticism is enough to dazzle. The idea of jungle becomes real through detailed work on the vertical composition and relationships of scale.

05. 金属网蚕蛹的原型
06. 研究草图，雕塑形体设计，昆廷·卡海尔
07. 钟罩凹室的内部
08. 构思草图
09. 神奇花园透视图
10. 温室花园构思草图

05. Prototype of the cocoon weaving
06. Study sketch, sculpture motifs, Quentin Garel
07. Inside the alcoves
08. Research sketch
09. Perspective image of the Garden of Marvels
10. Sketch for the Winter Garden

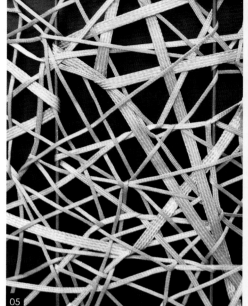
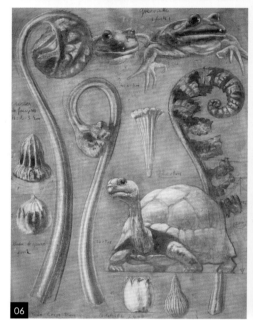

因此出现了一个向空中发展的花园，其图案性强并且形状细长，由树状蕨类植物、椰子树和铁树组成。一些树木也加入到这片植物群中，例如带有粉红色花、横向生长的异叶黄钟木。

在植物群中央，出现了三个由金属网编制而成的蚕蛹，这些独特的大型作品创造了壮观的美景。为了塑造这些景观效果，穆塔彼利斯事务所定期和两个艺术家进行合作：昆廷·卡海尔负责生命世界的造型雕塑，斯蒂芬妮·布提尔则负责编织艺术。

An aerial presence is established, rangy and graphic, composed of arborescent ferns, cabbage trees and cordylines. This vegetation is joined by trees, such as the *Tabebuia heterophylla* with its pink flowers and horizontal growth.

At the heart of this vegetation three cocoons rise up, monumental one-off pieces made of woven metal that create a fantasia. To make them, Mutabilis brought together two artists who work regularly with the agency: Quentin Garel for his sculptures of the living world and Stéphanie Buttier for her art of weaving.

04 私人花园与景观空间 Private gardens and landscape spaces

11-13. 一个植物繁茂的花园带人进入迷幻世界
14. 悬空蚕蛹的内部

11-13. A luxuriant garden which plunges us into a charming universe
14. Inside the cocoons

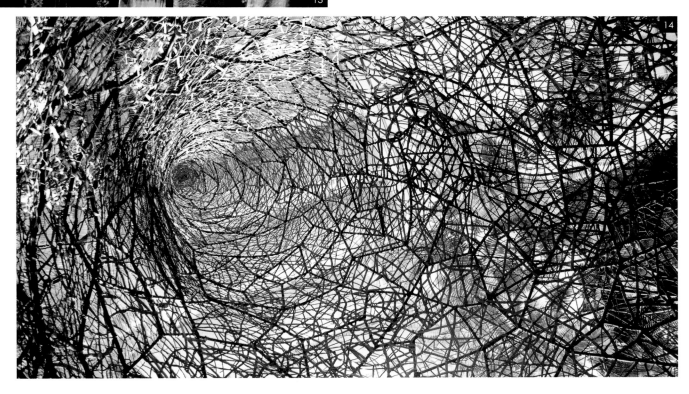

04 私人花园与景观空间 Private gardens and landscape spaces

01

陶艺花园
Potters' Gardens

NIEZ STUDIO

地点：法国萨尔格米纳
完工日期：2009
面积：1.5 ha
业主：萨尔格米纳镇政府
图片版权：Niez Studio

Location: Sarreguemines, France
Completion date: 2009
Area: 1.5 ha
Client: Sarreguemines Town Council
Photo credits: Niez Studio

01. 石磨花园里的狐尾百合
02. 废墟花园
03. 春天的石磨花园

01. The *Eremurus* 'Cléopatra' of the Millstone Garden
02. The Maze of Ruins garden
03. The Millstone Garden in spring

位于比利磨坊的陶艺花园，原先是生产用于陶艺的陶土工厂，如今成为陶艺技术博物馆的所在地。这个由几个主题空间构成的漫游场所，其构思与此工业场所的过往紧密相连。

设计的目的是通过空间整治和不同植物种类，唤起一系列颜色的变化和陶土的生产，以保存此地的"场所精神"。这个花园被特别构思成为一个在丰富植物世界中的漫步场所。

游客最先来到的是牡丹露台，其上种植着东方牡丹以及一款非常典型的当地牡丹"摩泽尔"。接下来是废墟迷宫，这是一处手工工场的遗迹，如今保留了现状，并逐渐成为攀缘植物的家园。然后是石磨花园，废弃的石磨与种植着一片壮观的火焰色独尾草丛融合在一起。阔叶植物园隐藏在两片石笼墙之间，由一条充斥着人工薄雾的小溪所组成，并且庇护着一片适合潮湿生态环境的植物。一条小路盘桓着观景台周围，迂回而上将人引领到一个可以远望整个景观的制高点。最后出现的是一片自发性小树林，在遗留于现场的陶瓷碎片层上，生长着质朴的乡野植物，这些植物与其伸出地面的树根形成了犹如雕塑作品的景观。

On the site of the Blies mill, formerly used for making pottery clay, the Potters' garden today forms the setting for the Museum of Pottery Techniques. Made up of several themed areas to be wandered through, its design was influenced by the site's industrial past.

The aim was to preserve the genius loci by using the layout and the plants to evoke the alchemy of colours and the making of the pottery. But above all this garden was conceived as a walk through the heart of a rich plant world.

First of all comes the terrace of Tree Peonies and its Oriental peonies cheek by jowl with the very local peony "La Moselle". Next it's the Maze of Ruins, the vestiges of the factory kept as they were found and progressively conquered by climbing plants. The Millstone Garden follows, where abandoned millstones are scattered among islands planted with the spectacular flame coloured Eremurus. The Big Leaf Garden, hidden between two gabion walls, is made up of a stream shrouded in artificial mists and hosts a wet environment flora. Around the Belvedere unfolds a path opening up on a viewpoint that embraces the surrounding landscape, and finally there appears the Spontaneous Smallwood in which rustic trees flourish on the strata of abandoned broken pottery, thus forming sculptures with their roots emerging from the soil.

04. 通往小溪流的路径
05. 废墟花园
06. 春天花朵绽放的樱桃树

04. Pathway to the stream
05. The Maze of Ruins garden
06. Cherry trees in flower in spring

04 私人花园与景观空间 Private gardens and landscape spaces

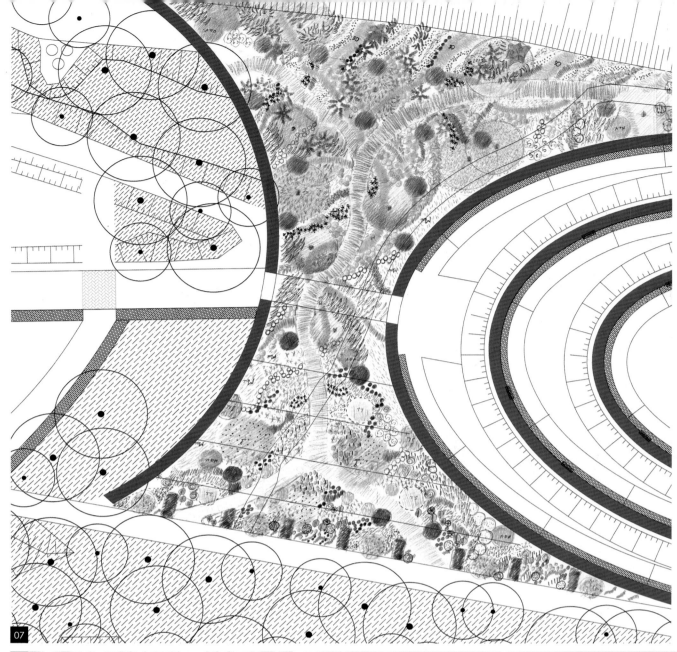

07. 阔叶植物园平面图
08. 阔叶植物：大叶子
09. 小溪薄雾中隐约可见的吊桥
10. 阔叶植物园剖面图

07. Planting plan for the leafy plant beds
08. Leafy plants: the *Rodgersia tabularis*
09. The footbridge over the misty brook
10. Cross-section of the leafy plants

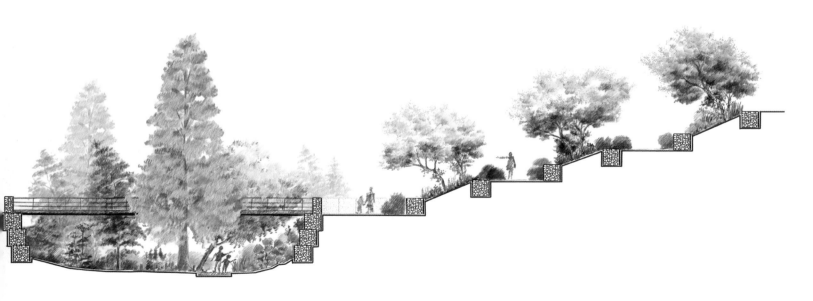

04 私人花园与景观空间 Private gardens and landscape spaces

11. 秋季花园景致
12. 浅溪过道秋季景致
13. 浅溪过道春季景致
14. 浅溪过道夏季景致

11. The garden in autumn
12. The stepping stones in autumn
13. The stepping stones in spring
14. The stepping stones in summer

04 私人花园与景观空间 Private gardens and landscape spaces

奥德赛花园 2000
Odyssey 2000 Gardens

PÉNA & PÉÑA PAYSAGISTES

地点：法国南泰尔
完工日期：2011
面积：1.63 ha
业主：法国巴黎银行房地产机构
图片版权：Christine & Michel Péna

Location: Nanterre, France
Completion date: 2011
Area: 1.63 ha
Client: BNP Paribas Immobilier
Photo credits: Christine & Michel Péna

01. 鸢尾花园和其间的弯曲水渠
02. 从建筑物天桥俯视鸢尾花园和弯曲水渠：将人引到公园
03. 鸢尾花园中的水渠与过桥

01. The iris garden and its canal
02. The iris garden and its canal seen from a footbridge: introduction to the park
03. Canal crossing in the iris garden, under the deck

奥德赛花园位于一块以工业活动为主的基地上，具有诸多限制并且相当缺乏绿化。由于接近塞纳河和船坞，因此水的存在透过特别的光线质量、河岸植被和河流气味，特别被强调出来。此项目设计目标也在于扭转今日稍显负面的基地形象：船坞以死巷的形式在设计地块之前停止，并带着不甚吸引人的色彩与外貌。

由于水具有缓和情绪的功能，在此建造一个以水景为主题的园地成为景观设计构思的重点。镜般的水面反射着天气的阴晴，呈现出变化多端的面貌，时而澄清明晰，时而被几许疾风所撩动。水景形态也呈现多样性格：在主入口与公园之间以沟渠形式来作为转化空间，在餐厅前面是植物满布的水池，在船坞前端则以镜光池面来吸引人们的注意，一座大型浮桥塑造出退缩的效果，避免人们与船坞混水的直接接触。

The Odyssey 2000 gardens form part of a mainly industrial site with a dominance of hard surfaces and many constraints. The presence of water, thanks to the river dock and the proximity of the Seine, is felt by a particular quality to the light, a riverside vegetation and river smells. There was a desire to overturn its image, which is today rather negative: the dock ends in a cul-de-sac in front of the land area, with a colour and surface that are far from attractive.

Creating a mainly aquatic park seemed interesting because of the way that water always calms the spirit. As a mirror, it reflects the moods of the weather that transform it into a changing surface, sometimes limpid and reflective, sometimes troubled by gusts of wind... The landscape architects imagined it taking on different characters – harnessed and channelled to ensure the transition between the main entrance and the park, a planted lake in front of the restaurants, an ornamental lake to catch the eye in front of the port, a large pontoon giving an effect of stepping back and avoiding immediate contact with the murky waters.

04. 凉亭平台、镜光池面、拖船小径和船坞
05. 船坞上的观景平台
06. 沿着船坞的镜光池面、观景平台和堤岸
07. 绿化剧场的高处
08. 日式踏脚石在草地与水池之间产生变化

04. The snack bar terrace, the mirror pool, the haulage path and the inner harbour
05. Lookout point over the harbour
06. The mirror pool, lookout point and quay, along the inner harbour
07. Looking down on the open-air theatre
08. Variations on the stepping stone, on grass and water

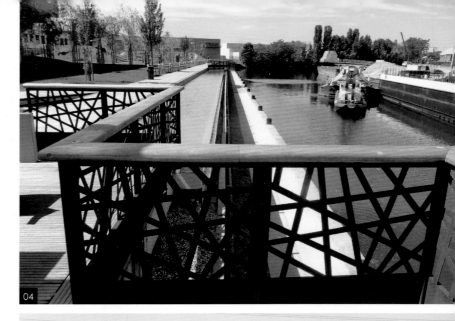

公园犹如展现新建筑的珠宝盒，而一群花园则穿越建筑中庭，并四处点缀着小平台。一座由污染土壤堆积而成的小山丘，由于移除费用过于昂贵，于是被塑造成绿化剧场，提供眺望船坞远端的视野，同时其备有座椅的缓坡也成为人们野餐的场所。为了回应高品质环境的要求，基地内的所有土地都具有一定的透水性：草地上的铺石、稳固的沙土地面以及木质铺板等，都允许雨水与水流能够很快渗入地下。一系列湿生植物为水岸带来绿化：覆地植物包括禾本植物、芦苇、鸢尾、千屈草和其他的排草属植物；树木包括梣木、杨树、柳树、橡树和樱桃树；灌木则采用低矮和线状的柳树。

The park serves as a setting for the new building, itself crossed by gardens and inset with patios. A hillock of polluted earth, too costly to remove, has been remodelled into an open-air theatre, offering a view over the port, its gentle slopes equipped with seats encouraging picnics. To respond to the high environmental quality that the site deserved no ground has been waterproofed: slabs on grass, stabilised sandy surfaces and wooden decking ensure a perfect permeability for rain and runoff water. As for the plants, the watersides have been given a variety of water-loving plants: grasses, reeds, irises, loosestrifes and other Lysimachea for the ground cover, and ash, birch, willow, oak, pine and cherry for the trees, with coppiced and row-planted willow for the shrubs.

04 私人花园与景观空间 Private gardens and landscape spaces

09. 面对着餐厅、通往半岛的日式踏脚石
10、11. 在水池边缘的餐厅露天座
12、13. 绿化中庭
14. 水池和餐厅露天座

09. Stepping stones towards the peninsula, facing the restaurant terraces
10-11. The restaurant terraces overlooking the pool
12-13. The green patio
14. The pool and the restaurant terraces

04 私人花园与景观空间 Private gardens and landscape spaces

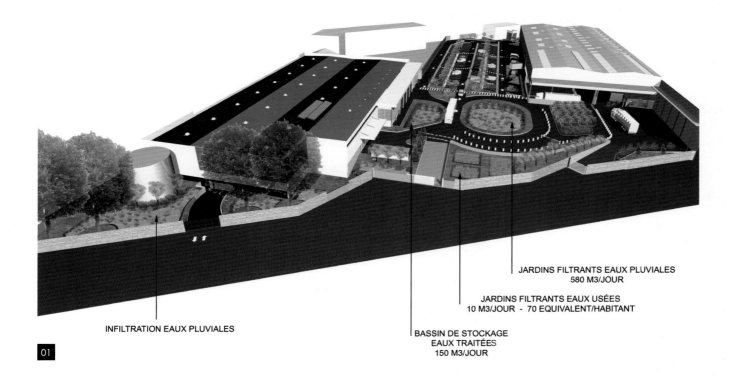

01

Gardens for La Plateforme du Bâtiment
建筑物平台企业花园
PHYTORESTORE / THIERRY JACQUET

地点：法国欧贝维利耶
完工日期：2010
面积：2.4 ha
业主：Point P / 圣戈班集团
合作设计师：Agence d'architecture Norbert Brail
图片版权：Phytorestore & Agence d'architecture Norbert Brail (n°01), Phytorestore / Thierry Jacquet (n°02-09)

Location: Aubervilliers, France
Completion date: 2010
Area: 2,4 ha
Client: Point P / Groupe Saint Gobain
Co-project manager: Agence d'architecture Norbert Brail
Image credits: Phytorestore & Agence d'architecture Norbert Brail (n°01), Phytorestore / Thierry Jacquet (n°02-09),

01. 方案全景剖面透视图
02. 储存雨水的草沟
03. 雨水渗透区

01. Overall plan
02. Stormwater harvesting swale
03. Stormwater infiltration zone

建筑物平台（La Plateforme du bâtiment）企业用地位于巴黎城市化密度极高的城门区域，其景观整治仅占用基地20%的面积，却能够使废水与雨水获得就地处理与循环，而完全不需外排。项目的优点之一也在于对处理过后的废水的重新使用。经收集的雨水立即通过有机过滤而储存在景观化的水塘里，以作为卫生间用水；处理过的废水则用来灌溉4500平方米的绿地以及分散在基地的150棵新植的树木。

这个企业仓库位于圣马丁运河附近，其整治项目占地2.4公顷。项目方案企图保存最大量的绿地，因此不仅一部分的旧建筑遭到拆除，新建筑也后退建造。所有即存的大型梧桐全部被保留下来，并且经过修剪以避免影响新店面的建造。由于项目遵守HQE（高环境质量）准则而建造，因此仅仅种植不会造成过敏效果或几率极低的树种，例如桤木、欧洲鹅耳枥、白蜡树和一些极为特别的冷杉。同时也仅种植质朴的草原，而避免使用传统草坪。

The landscape project for La Plateforme du Bâtiment builders' merchant, situated in a dense urban fabric just outside central Paris, shows how a landscaped development using 20% of the surface area of the plot can enable the management of all the waste water and stormwater (50-year floods). The project is also notable for its recycling of all the treated water. The stormwater, which passes immediately through organic filters (peat), is stocked in a landscaped pond and reused for the toilets. The treated waste water serves for the underground irrigation of 4,500 m² of green spaces and 150 new trees, apportioned over the whole site.

The development project concerns 2.4 hectares of a commercial depot near the Saint-Martin navigational canal. It prioritised the preservation of the largest area of green spaces possible. Part of the old construction was demolished and the new building set back in position. All the large old plane trees have been preserved and pruned so as not to block the construction of a new store. As the project embraces a High Environmental Quality (HQE) approach, only tree species that have no or little allergenic potential, such as alder, hornbeam, ash and certain species of pine, have been used. The choice of rustic meadows avoids the use of classic lawns.

04. 过滤雨水的花园
05. 过滤废水的花园
06. 经过处理之废水的渗透区
07. 雨水过滤剖面图
08. 废水处理剖面图
09. 停车场前的野生草原

04. Stormwater filtering gardens
05. Wastewater filtering gardens
06. Treated water infiltration zone
07. Section of the stormwater filter
08. Section of the treatment of raw wastewater
09. Wild meadow in front of the car park

方案的困难之一在于说服甲方接受设置一定深度的景观草沟。为了避免形成荒原景象，沟岸以柴笼堤坝来建造，以为基地塑造一定的景观结构。此外，为了能够阅读基地的水纹，让水的存在明显可识别，景观设计师特别选择了典型的湿地植物，例如菖蒲、莎草和灯心草，以塑造明显的湿地氛围。为了协助鸟禽驻入，此地还特别设置了一些鸟笼和茂密的芦苇丛。其他生物随后驻入证实了方案的成功。

One of the next difficulties was to find a way to incorporate deep swales into the landscape. To avoid too wild an appearance, the embankments have been treated with timber piling, enabling the structuring of the landscape of the site. To make the water course legible and thus retain a visible presence of water, typical wetland plants like irises, sedges and rushes have been chosen to strengthen the ambiances of these wetland zones. Nesting boxes and dense reed beds were installed to encourage birds. A study of species sightings shows that these measures have been successful.

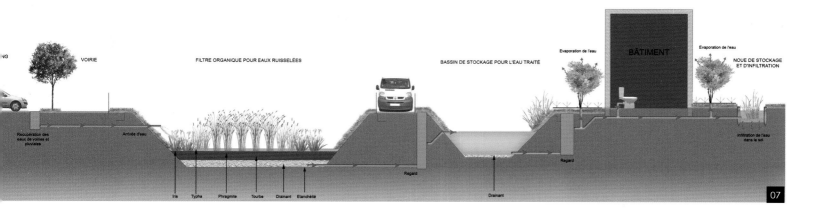

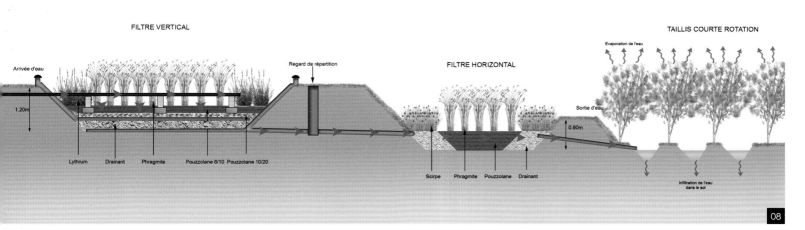

04 私人花园与景观空间 Private gardens and landscape spaces

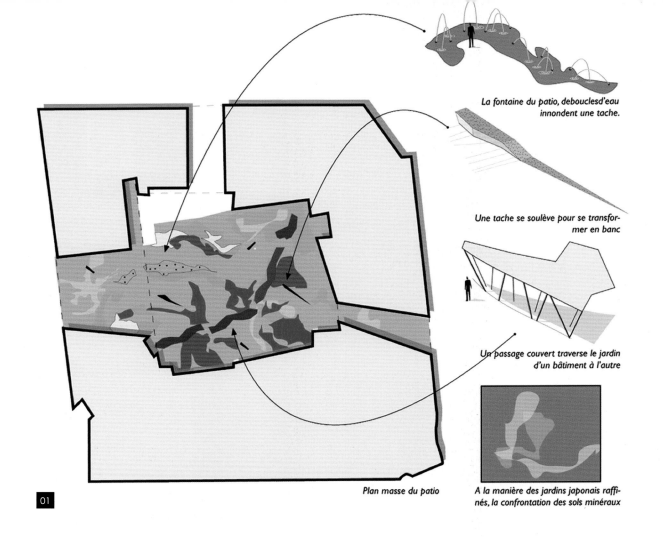

01

"Jackson Pollock" Office Garden
"杰克逊·波洛克"办公楼花园

LAURE QUONIAM

地点：法国圣丹尼
完工日期：2011
面积：1 600 m²
业主：法国93地区共同平原发展组织
照片版权：Marine Saiah

Location: Saint-Denis, France
Completion date: 2011
Area: 1 600 m²
Client: Plaine Commune Développement 93
Photo credits: Marine Saiah

01. 整体平面配置图
02. 花园的创作灵感来自于杰克逊·波洛克的画作
03. 花园融入建筑体的等角透视
04. 依照画作精神而设计的通风格栅

01. Master plan
02. The garden's source of inspiration – a painting by Jackson Pollock
03. Axonometric drawing of the future garden inserted into the building
04. The ventillation grille adapted in the spirit of the painting

本项目办公楼位于接近法国运动场和巴黎城门的新建龙迪-佩雷耶勒街区之内，其1600平方米的建筑中庭设计是依据杰克逊·波洛克的画作所作的景观诠释。从各楼层的窗户向中庭眺望，都可看到这个作品。

这个以花园体现画作的决定使得此方案不论在平面构图或者材质选用上都呈现令人惊叹的效果。弧形线条与柔和形态使画作得以融入建筑物的框架中。此设计构思特别是为了让人从高处观赏景观，犹如在一定距离之外欣赏画作。在此，景观师刻意选择低矮的植物种类，并且让花园的色调随着季节的演进而产生变化。一道斜线小径通往两个办公室主要入口，而餐厅则直接面向花园。

植物的体量重新表达出画作的图面效果，地面的材质也遵循着其线条形态。杜鹃和所采用的覆地植物都展现出鲜艳的色彩：红、黄、橙等，并彼此互相结合。而地面则是由石英溶质凝固而成的碎石铺设而成。植物的缤纷色彩与地面色调相互重叠塑造出绚丽效果。黑色水泥制成的椅凳从地面窜出，仿佛画家在画布上划上了一刀。这三张椅凳是特别为此中庭花园所设计的。

Situated in one of the office buildings of the new neighbourhood of Landy Pleyel, just outside Paris and near to the Stade de France, the patio is an interpretation of a painting by Jackson Pollock. Covering 1,600 m², it can be viewed from above from all the surrounding windows.

The choice of reproducing this painting opened the way for surprising effects, both in terms of the line plan and in the composition of its materials. The curved, supple forms allowed the figure in the painting to be introduced into the structured framework of the building. It was essentially designed to be seen from the floors above, just as you look at a painted surface from a distance. The garden changes constantly as the seasons progress, like an animated film. The plant species were deliberately selected for their low-lying growth. A diagonal path leads to the two main entrances, and the restaurant opens onto the patio at the same level.

Volume in the planting has been used to reproduce the visual effects of the painting. The materials on the ground echo the morphological forms of its lines. Azaleas and ground-covering plants in bright colours – red, yellow, orange – reflect the pigments in the picture. The colourful plant masses are superimposed on the hard surface material, made from a quartz gravel that is poured then solidified. The black concrete benches rise up from the ground as if the painter had slashed the canvas with his cutter. Made in three pieces that are then reassembled, they were designed specifically for this patio.

05. 地面视角花园景观
06. 画作的笔触转化为植物丛
07、08. 从办公楼俯视花园
09. 水泥椅凳细部

05. Ground level view of the garden
06. Splashes of colour transposed into plant masses
07-08. View from the upper floors
09. Detail of the concrete bench

04 私人花园与景观空间 Private gardens and landscape spaces

Pierre-Joël Bonté School for Building Skills
皮埃尔-乔尔·邦德建筑职业高中

LAURE QUONIAM

地点：法国里永-沃尔维克
完工日期：2009
面积：10 ha
业主：奥弗涅地区议会
照片版权：Marine Saiah

Location: Riom-Volvic, France
Completion date: 2009
Area: 10 ha
Client: Auvergne Reginal Council
Photo credits: Marine Saiah

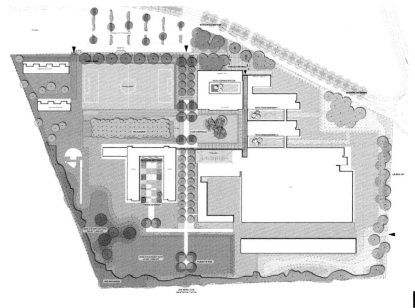

01. 从教学楼看向中庭的景观
02. 整体平面配置图：对水流的控制
03. 基地边缘的排水净化系统
04. 从公园看向教学楼

01. View of the patio from the teaching building
02. Site plan – control of water runoff
03. Water purification system on the periphery
04. View from the park

一所旨在提升营建业地位的新职业高中聚集了1230名学生，其建筑物展示出木质建材的完备性能。学校对于不同建筑和公园之间的连接予以细心规划，使学生的流动路线穿越诸多氛围明朗的生活场所。面积10公顷的公园空间围绕着两条主要轴线而组织，南北向轴线提供了眺望远处山脉的景观，东西向轴线则串联一系列与营建有关的空间和各种运动场所。

入口前庭扮演接待的角色，中央林荫道则是校园空间的主要骨架，并且组织着所有交通流线的分配。树身修长的帚状千金榆在建筑物之间呈直线排列，形成屏幕。滚球游戏场是一个下陷的长方形大型草坪，其尺寸使它成为建筑配置的中心场所，周围的大型通道连接着餐厅、宿舍、资料信息中心和教室。教学楼与宿舍楼皆设置有氛围较为私秘的中庭空间。

学生运动场分布在建筑物四周的草地上，不同的排水净化系统（防洪草沟、油漆分离器、蓄水池、过滤植物）以及一个回收所有水的传统地下化系统则设置在校园边缘地带。

The 1,230 building skills students have been brought together in a new school that celebrates excellence in the building professions. It also demonstrates the efficiency and performance of wood. Care has been taken to organise the links between the different buildings and the park. In the course of their day the students cross several living spaces with atmosphere and light. The park covers 10 hectares and is organised around two compositional axes. The north-south axis makes the most of the distant views towards the Puys mountain range. The east-west axis is a series of sub-spaces linked to the life of the building and to the many sports activities.

The entrance forecourt is designed to be welcoming. A central walk forms the backbone of the project, from which all the other paths extend. A row of pyramid hornbeams with their slender forms runs between the buildings. The "bowling green", a large manicured lawn, stands out at the heart of the architectural project, where a wide passageway leads to the restaurant, the dormitories, the library and classrooms. The patios are on a more intimate scale and are accessed from the classroom and dormitory buildings.

Sports grounds are available to the students in the fields that surround the building. On the periphery of the school, an alternative water purification system (swales, hydrocarbon separator, retention pond, filtering plants) and a traditional underground system harvest all the water from the site.

05. 东北轴线道路
06. 学校入口前庭
07. 面对食堂的中央广场
08. 行政楼中庭
09. 宿舍中庭

05. North-east avenue
06. School entrance – forecourt
07. Central square facing the canteen
08. Patio of the administration block
09. Dormitory patio

04 私人花园与景观空间 Private gardens and landscape spaces

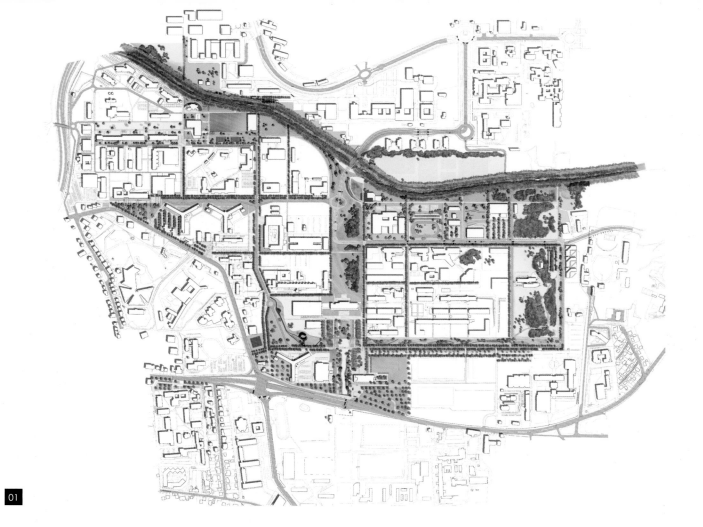

宏格耶大学校园
Rangueil Campus

URBICUS

地点：法国图卢兹
完工日期：自2012年起施工
面积：152 ha
业主：SGE, PRES
图片版权：Urbicus (n°01, 02, 04, 05), Cube (n°03, 06-08)

Location: Toulouse, France
Completion date: ongoing since 2012
Area: 152 ha
Client: SGE, PRES
Image credits: Urbicus (n°01, 02, 04, 05), Cube (n°03, 06-08)

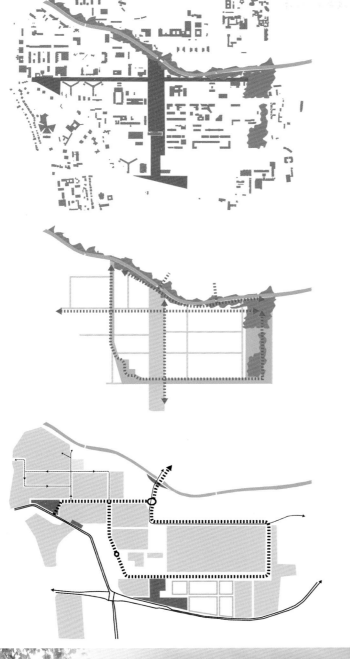

01. 宏格耶校园为图卢兹城市提供了一个崭新的公园
02. 此方案对校园的空间与使用功能进行了清晰的分级处理，以塑造出最佳的城市组织与场所识别性
03. 子午运河在其经过的两岸形成了一个线性公园，并且以各种开口和衔接方式来与校园产生联系

01. Rangueil campus gives Toulouse a new park
02. A hierarchy of spaces and uses for a better understanding of the space and urban organisation
03. Through openings and links with the campus, the Canal du Midi gives rise to a "linear park"

此方案促进了图卢兹欧洲国际大学的科学研究与教学质量的提升，在宏格耶街区所汇集的多种接待功能、其学生活动的质量、与城市的融合以及对社会和创新所持有的开放精神等层面，透过景观的整治而使得街区更具有吸引力。

这个日趋老化的校园在整治之后转变成一个交通便利的舒适街区，其中机动车的数量非常的低，几个结构性的公共空间都带着图卢兹城市的特色，每栋建筑物皆沉浸在自然公园之中，而学生与教授们都因为能够在此居住和工作而感到荣幸之至。校园里的阶梯式大讲堂、实验室、教室都与户外经过整治的大自然交相呼应。

一个位于街区入口处、充满林荫的大型露天绿化剧场成为人们聚集、相会与交流的地标性场所，为校园带来崭新的意象和实用方式。在街区的各个入口皆设有大型广场，以迎接公共交通的抵达。一条南北轴线和一条东西轴线将整个大学空间组织了起来，并且使其与城市产生联结。此外，一个由植物与水径所组成的外围绿化带提供了休闲与运动的场所。在基地和大城乡尺度上都构成主要公共空间的子午运河，在其两岸形成了一个线性公园，街区的建筑立面必须重新转向这个宜人的绿化空间。

The project plays a part in the making of scientific and educational excellence at the European and International University of Toulouse. It adds value to the Rangueil landscape in order to make it attractive through its welcomeness to all, through the quality of its student life, through its integration into the city, and through its openness to society and innovation.

This ageing campus will become a well-served and comfortable neighbourhood with few cars, where the structuring public spaces have an identity specific to Toulouse, and where each building is presented in an inhabited natural park where students and teachers are proud to live and work. Amphitheatres, laboratories and classrooms find an exterior continuation in an enhanced version of the existing nature.

A large open-air amphitheatre, a shady entrance to the neighbourhood, creates a space for gathering, a landmark and a meeting place that strongly rejuvenates the image and the life of the campus. On arrival at the neighbourhood, wide esplanades welcome people arriving by public transport. A "cardo" and a "decumanus" – the main north-south and east-west streets of a Roman city – are written into and attach themselves to the university in the city. A green belt of nature and water is a place for relaxation and sports pursuits. The Canal du Midi, a major public site and important for the urban area, generates a "linear park" on its banks, to which the facades of the neighbourhood will turn.

04. 充满动感与活力的阶梯空间为学生们提供了宜人的户外生活场所
05. 沿着子午运河的阶梯空间铺设着舒适的草地
06. "大型露天绿化剧场"为校园带来崭新的意象
07. 被大型广场花园所衬托的校园入口之一
08. 一个随着季节与雨量而变化的下沉式花园

04. Terracing with seats designed for student life
05. Grassy amphitheatre terraces along the Canal du Midi
06. The "large green amphitheatre" gives the campus a new image
07. A large garden square marking one of the entrances to the campus
08. A sunken garden planted according to the seasons and the amount of rain

04 私人花园与景观空间 Private gardens and landscape spaces

附录
设计师索引

Annex
The Designers

AGENCE APS
Hubert Guichard / Jean-Louis Knidel / Gilles Ottou

31 Grande Rue 26000 Valence
T +33 4 75 78 53 53
F +33 4 75 78 53 50
agenceaps@me.com
www.agenceaps.com

pp.20-25, pp.94-99, pp.192-197, pp.384-389

AGENCE TER
Olivier Philippe / Henri Bava / Michel Hoessler

18-20 rue du Faubourg du Temple 75011 Paris
T +33 1 43 14 34 00
F +33 1 43 38 13 03
contact@agenceter.com
www.agenceter.com

pp.100-103, pp.198-201, pp.202-205

AGENCE TERRITOIRES
Étienne Voiriot / Philippe Convercey / Franck Mathé

22 rue Mégevand 25000 Besançon
T +33 3 81 82 06 66
F +33 3 81 82 08 09
info@territoirespaysagistes.com
www.territoirespaysagistes.com

pp.104-109

ARTE CHARPENTIER
Équipe Paysage :
Nathalie Leroy / Élodie Ledru / Laurène Moraglia / Lara Pilotto

8 rue du Sentier 75002 Paris
T +33 1 55 04 13 55
F +33 1 55 04 13 13
contact@arte-charpentier.com
www.arte-charpentier.com

pp.206-209, pp.390-395

ATELIER DE L'ÎLE
Bernard Cavalié

3 rue Dagorno 75012 Paris
T +33 1 48 06 22 00
F +33 1 48 06 91 75
paris.atile@atile.fr
www.atile.fr

pp.234-239, pp.240-245

ATELIER RUELLE
Gérard Pénot

5 rue d'Alsace 75010 Paris
T +33 1 55 04 89 99
F +33 1 55 04 89 69
atelierparis@atelier-ruelle.fr
www.atelier-ruelle.fr

pp.224-227, pp.228-233

ATELIER VILLES & PAYSAGES
Pierre-Michel Delpeuch / Jean-Marc Bouillon

170 avenue Thiers 69455 Lyon Cedex 06
T +33 4 37 72 45 01
F +33 4 37 72 27 11
lyon@villesetpaysages.fr
www.villesetpaysages.fr

pp.26-31, pp.210-215, pp.216-223

ATELIER DE PAYSAGES BRUEL-DELMAR
Anne-Sylvie Bruel / Christophe Delmar

40 rue Sedaine 75011 Paris
T +33 1 47 00 00 51
F +33 1 47 00 13 51
contact@brueldelmar.fr
www.brueldelmar.fr

pp.32-37, pp.246-253, pp.254-259, pp.260-271

SOPHIE BARBAUX

24ter impasse Berrin 13010 Marseille
T +33 6 08 64 02 84
s.barbaux@orange.fr
www.sophie-barbaux.odexpo.com

pp.396-397, pp.398-401

LOUIS BENECH

4 Cité Saint Chaumont 75019 Paris
T +33 1 42 01 04 00
F +33 1 42 01 01 05
agence@louisbenech.com
www.louisbenech.com

pp.402-405, pp.406-409

MICHEL CORAJOUD

23 rue Sébastien Mercier 75015 Paris
T +33 1 44 37 06 60
michel.corajoud@wanadoo.fr
www.corajoudmichel.nerim.net

pp.110-115

COULON LEBLANC & ASSOCIÉS
Jacques Coulon / Linda Leblanc

89 rue du Faubourg Saint Antoine
75011 Paris
T +33 1 40 09 15 11
F +33 1 40 09 18 22
atelier.jacquescoulon@orange.fr
www.coulon-leblanc.fr

pp.38-41, pp.272-275

MICHEL DESVIGNE PAYSAGISTE

23 rue du Renard 75004 Paris
T +33 1 44 61 98 61
F +33 1 44 61 98 60
contact@micheldesvigne.com
www.micheldesvigne.com

pp.42-45, pp.46-49, pp.116-121, pp.276-281

IN SITU
Emmanuel Jalbert

8 quai Saint Vincent 69001 Lyon
T +33 4 72 07 06 24
contact@in-situ.fr
www.in-situ.fr

pp.126-133, pp.134-137, pp.296-311, pp.312-315

DIGITALE PAYSAGE
Agnès Daval / Bruno Steiner

39 rue de l'École 67330 Imbsheim
T/F +33 3 88 71 37 68
contact@digitalepaysage.com
www.digitalepaysage.com

pp.50-53, pp.282-287

L'ANTON & ASSOCIÉS
Jean-Marc L'Anton

31 avenue Laplace 94110 Arcueil
T +33 1 49 12 10 90
F +33 1 49 12 10 86
agence.lanton@wanadoo.fr
www.agence.lanton.com

pp.316-323, pp.324-329, pp.428-431

DVA PAYSAGISTES
Olivier Damée

12 & 12bis rue Mélingue 75019 Paris
T +33 1 42 06 38 60
F +33 1 42 06 38 69
d.v.a@dvapaysages.com
www.dvapaysages.com

pp.54-57

FLORENCE MERCIER PAYSAGISTE

85 rue Mouffetard 75005 Paris
T +33 1 44 08 80 25
F +33 1 43 31 60 00
contact@fmpaysage.fr
www.fmpaysage.fr

pp.64-69, pp.330-333, pp.432-435

MAISON ÉDOUARD FRANÇOIS

136 rue Falguière 75015 Paris
T +33 1 45 67 88 87
F +33 1 45 67 51 45
maison@edouardfrancois.com
www.edouardfrancois.com

pp.410-415

CATHERINE MOSBACH

81 rue des Poissonniers 75018 Paris
T +33 1 53 38 49 99
F +33 1 42 41 22 10
mp.mosbach@mosbach.fr
www.mosbach.fr

pp.70-75, pp.76-81, pp.138-143, pp.144-147

AGENCE HORIZONS
Jérôme Mazas

61 boulevard Bompard 13007 Marseille
T +33 4 91 46 38 60
F +33 4 91 46 31 22
contact@horizons-paysages.fr
www.horizons-paysages.fr

pp.58-63, pp.416-421, pp.422-427

MUTABILIS
Juliette Bailly-Maître / Ronan Gallais

4 passage Courtois 75011 Paris
T +33 1 43 48 61 33
F +33 1 43 48 93 23
mutabilis.paysage@wanadooo.fr
www.mutabilis-paysage.com

pp.82-87, pp.334-337, pp.436-441

HYL
Pascale Hannetel / Arnaud Yver

90 rue du Chemin Vert 75011 Paris
T +33 1 49 29 93 23
F +33 1 49 29 45 61
paysage@hyl.fr
www.hyl.fr

pp.122-125, pp.288-291, pp.292-295

NIEZ STUDIO
Philippe Niez

18 rue Yves Toudic 75010 Paris
T +33 1 43 66 58 81
contact@philippeniez.com
www.philippeniez.com

pp.442-449

OLM
Philippe Coignet

156 rue Oberkampf 75011 Paris
T +33 1 42 06 44 51
F +33 1 42 06 64 27
info@o-l-m.net
www.o-l-m.net

pp.88-91, pp.148-149, pp.338-341

SAVART PAYSAGE
Marc Soucat

23 rue des Vertus
51000 Châlons-en-Champagne
T/F +33 3 26 26 99 71
savart.paysage@orange.fr
www.savart-paysage.com

pp.354-357

ATELIER JACQUELINE OSTY & ASSOCIÉS

77 rue de Charonne 75011 Paris
T +33 1 43 48 63 84
F +33 1 43 67 16 75
atelier@osty.fr

pp.150-155, pp.156-161

TN PLUS
Andras Jambor / Jean-Christophe Nani / Bruno Tanant

30 boulevard Richard-Lenoir 75011 Paris
T +33 1 43 55 42 07
agence@tnplus.fr
www.tnplus.fr

pp.180-183, pp.358-361, pp.362-367, pp.368-371

PÉNA & PEÑA PAYSAGISTES
Christine & Michel Péna

15 rue Jean Fautrier 75013 Paris
T +33 1 45 70 00 80
F +33 1 45 70 72 66
contact@penapaysages.com
www.penapaysages.com

pp.168-173, pp.342-345, pp.346-349, pp.450-455

TRÉVELO & VIGER-KOHLER
Pierre-Alain Trévelo / Antoine Viger-Kohler

23 rue Olivier Métra 75020 Paris
T +33 1 47 00 04 62
F +33 1 47 00 08 85
agence@tvk.fr
www.tvk.fr

pp.372-375

PHYTORESTORE
Thierry Jacquet

146 boulevard de Charonne 75020 Paris
T +33 1 43 72 38 00
F +33 1 43 72 38 07
info@phytorestore.com
www.phytorestore.com

pp.162-167, pp.456-459

URBICUS
Jean-Marc Gaulier

3 rue Edme Frémy 78000 Versailles
T +33 1 39 53 14 35
F +33 1 39 49 46 23
axp@urbicus.fr
www.urbicus.fr

pp.184-189, pp.376-381, pp.468-471

ALLAIN PROVOST

5 rue de Naples 78150 Rocquencourt
T +33 1 39 02 12 55
al.pro@orange.fr

pp.174-179

LAURE QUONIAM

6 rue des Tournelles 75004 Paris
T +33 1 46 33 77 25
agence@laurequoniam.com
www.laurequoniam.com

pp.350-353, pp.460-463, pp.464-467

图书在版编目（CIP）数据

现代景观设计 / 法国亦西文化编；邵雪梅，陈庶，简嘉玲译. — 沈阳：辽宁科学技术出版社，2013.8
ISBN 978-7-5381-8139-5

Ⅰ. ①现… Ⅱ. ①法… ②邵… ③陈… ④简… Ⅲ. ①景观设计－作品集－法国－现代 Ⅳ. ①TU986.2

中国版本图书馆CIP数据核字(2013)第151457号

出版发行：辽宁科学技术出版社
　　　　　（地址：沈阳市和平区十一纬路29号　邮编：110003）
印　刷　者：利丰雅高印刷（深圳）有限公司
经　销　者：各地新华书店
幅面尺寸：240mm×320mm
印　　张：59.5
插　　页：4
字　　数：11.5千字
印　　数：1～1200
出版时间：2013年 8 月第 1 版
印刷时间：2013年 8 月第 1 次印刷
责任编辑：陈慈良　隋　敏
封面设计：杨春玲
版式设计：亦西文化
责任校对：周　文
书　　号：ISBN 978-7-5381-8139-5
定　　价：498.00元

联系电话：024-23284360
邮购热线：024-23284502
E-mail：lnkjc@126.com
http://www.lnkj.com.cn
本书网址：www.lnkj.cn/uri.sh/8139